Springer Theses

Recognizing Outstanding Ph.D. Research

Aims and Scope

The series "Springer Theses" brings together a selection of the very best Ph.D. theses from around the world and across the physical sciences. Nominated and endorsed by two recognized specialists, each published volume has been selected for its scientific excellence and the high impact of its contents for the pertinent field of research. For greater accessibility to non-specialists, the published versions include an extended introduction, as well as a foreword by the student's supervisor explaining the special relevance of the work for the field. As a whole, the series will provide a valuable resource both for newcomers to the research fields described, and for other scientists seeking detailed background information on special questions. Finally, it provides an accredited documentation of the valuable contributions made by today's younger generation of scientists.

Theses are accepted into the series by invited nomination only and must fulfill all of the following criteria

- They must be written in good English.
- The topic should fall within the confines of Chemistry, Physics, Earth Sciences, Engineering and related interdisciplinary fields such as Materials, Nanoscience, Chemical Engineering, Complex Systems and Biophysics.
- The work reported in the thesis must represent a significant scientific advance.
- If the thesis includes previously published material, permission to reproduce this must be gained from the respective copyright holder.
- They must have been examined and passed during the 12 months prior to nomination.
- Each thesis should include a foreword by the supervisor outlining the significance of its content.
- The theses should have a clearly defined structure including an introduction accessible to scientists not expert in that particular field.

More information about this series at http://www.springer.com/series/8790

Sho Sugiura

Formulation of Statistical Mechanics Based on Thermal Pure Quantum States

Doctoral Thesis accepted by
The University of Tokyo, Tokyo, Japan

Springer

Author
Dr. Sho Sugiura
Department of Science
The University of Tokyo
Tokyo
Japan

Supervisor
Prof. Akira Shimizu
Department of Science
The University of Tokyo
Tokyo
Japan

ISSN 2190-5053 ISSN 2190-5061 (electronic)
Springer Theses
ISBN 978-981-10-1505-2 ISBN 978-981-10-1506-9 (eBook)
DOI 10.1007/978-981-10-1506-9

Library of Congress Control Number: 2017948625

Printed on acid-free paper

This Springer imprint is published by Springer Nature
The registered company is Springer Nature Singapore Pte Ltd.
The registered company address is: 152 Beach Road, #21-01/04 Gateway East, Singapore 189721, Singapore

Supervisor's Foreword

It is indeed a great pleasure for me to write this foreword to the Ph.D. Thesis of Sho Sugiura, whom I refer to as Sugiura-kun. He established a new formulation of quantum statistical mechanics, according to which all quantities of statistical mechanical interest including thermodynamic functions, are obtained from a single pure quantum state.

Twenty years ago, I had given a lecture on the statistical mechanics at the University of Tokyo. During the lecture, I stressed that almost all microstates with a macroscopically equal value of energy, are the equilibrium state specified by that energy value. This fact, recently called 'typicality', was said to be realized for classical systems a by Boltzmann and Gibbs. For the quantum systems, it was proved by von Neumann many years ago, and by many researchers in the last decade.

When I had lunch with Prof. Hukushima, a good colleague of mine, he said "You insist that typicality is the heart of statistical mechanics. If so, it should be able to formulate statistical mechanics using only a single microstate. Is it really possible?" This conversation motivated me to explore a new formulation with Sugiura-kun.

Although the typicality suggested that statistical mechanics could be formulated using a single microstate, it actually gave only 'mechanical variables,' such as energy and particle density. It seemed impossible to obtain 'genuine thermodynamic variables,' such as entropy and thermodynamic functions, from a single microstate. Furthermore, all the previous studies on the typicality were formulated in the microcanonical setup, which specifies equilibrium states by extensive variables such as energy. Microstates which represent equilibrium states in the canonical and grand-canonical setups were unknown. For these difficulty and limitation, it was impossible to formulate statistical mechanics using only a single microstate.

Sugiura-kun and coworkers have currently resolved these problems. They have constructed various pure quantum states, called thermal pure quantum (TPQ) states, which represent equilibrium states. The TPQ states include a microcanonical one, a canonical one, and a grand canonical one. Furthermore, they have also presented several formulas for obtaining 'genuine thermodynamic variables' from a single microstate. Therefore, their formulation can fully replace the conventional Gibbs formulation. This new formulation of quantum statistical mechanics is called the TPQ formulation, according to which all quantities of statistical mechanical interest are obtained from a single TPQ state.

This Ph.D. thesis by Sho Sugiura is the first book on the TPQ formulation. I believe that the reader will enjoy the thesis very much. I expect the thesis will also be useful for practical calculations because the TPQ formulation is shown to provide a good numerical method.

I am grateful to Sugiura-kun for many enlightening discussions, efforts, and considerations. I'd like to close my foreword by mentioning the delightful fact that this work has rewarded him with an opportunity to greet and chat with the 124th Emperor of Japan, when he was awarded the won the JSPS Ikushi prize.

Komaba, Tokyo, Japan
June 2017

Prof. Akira Shimizu

List of Papers Submitted for the Requirement of the Degree

1. S. Sugiura and A. Shimizu, *Thermal Pure Quantum States at Finite Temperature*, Phys. Rev. Lett. **108**, 240401 (2012).
2. S. Sugiura and A. Shimizu, *Canonical Thermal Pure Quantum State*, Phys. Rev. Lett. **111**, 010401 (2013).
3. S. Sugiura and A. Shimizu, *New Formulation Of Statistical Mechanics Using Thermal Pure Quantum States*, Kinki University Series on Quantum Computing: Volume 9 "Physics, Mathematics, and All that Quantum Jazz", 245–258 (2014).

Acknowledgements

I deeply acknowledge for Prof. Shimizu. Your continuous support and encouragement lead me here. When I imagine what if I have not been in your group, I even feel something like a fear that I may have done nothing in my Ph.D. days. I would also like to thank my parents and family for their continuous support. Although we have been facing many troubles, I am confident that we will overcome.

Contents

Chapter 1
Introduction to Thermal Pure Quantum State Formulation of Statistical Mechanics

Consider a glass of water placed gently on a table. The glass of water always relaxes to an equilibrium state, which are specified only by several physical quantities such as temperature and a volume. After waiting for a while, the glass of water still stays in the same equilibrium state. That is, when one measures quantities such as its temperature, volume, pressure, density, and so on, they always take the same value within the range of fluctuation. Thermodynamics is the theory which predicts properties of the equilibrium states in macroscopic systems. All thermodynamic predictions are obtained from thermodynamic functions such as entropy and free energy.

By contrast, if one can observe a glass of water in a microscopic scale, water molecules are moving on and on, and the microscopic state is changing continuously. This microscopic world is described by quantum (or Newtonian) mechanics. However, it is virtually impossible to know the microscopic state completely in the system which has Avogadro's numbers of particles. Moreover, even if we know it, we still do not understand why the macroscopic quantities such as the density and the pressure stay invariant, while the position and the momentum of each particle are changing.

Statistical mechanics is the theory which connects the gap between the microscopic world and the macroscopic one; it enable one to derive thermodynamical predictions from microscopic mechanics. To obtain the macroscopic predictions, we conventionally assume two principles in statistical mechanics. One is "the principle of equal a priori probability (PEPP)", and the other is "Boltzmann's entropy formula." Using these assumptions, statistical mechanics has achieved great success in every field in physics for over a century.

However, the justification of these principles has not yet been revealed. Especially, attempts to explain the PEPP from the microscopic mechanics have a long history [1–4]. Ergodicity, which tries to derive the PEPP from time evolution of a system, had been investigated thoroughly [2, 3, 5, 6] but has hardly succeeded in deriving it yet. Instead of the ergodicity, many statistical physicists nowadays believe that typicality among states [7–17] should be the foundation of statistical mechanics. The typicality is the property that almost all the realizable state can be regarded as

S. Sugiura, *Formulation of Statistical Mechanics Based on Thermal Pure Quantum States*, Springer Theses, DOI 10.1007/978-981-10-1506-9_1

the equilibrium state, and the ratio of non-equilibrium states are vanishingly small. Namely, we can almost certainly observe the equilibrium value when we prepare a single microstate and measure it. From this viewpoint, the PEPP, in which we take average over all the realizable states, is understood to be a very sufficient way to obtain this equilibrium value (or a way too much). Since the typicality can be the foundation of statistical mechanics, it has been proposed by many physicists, including Boltzmann, von Neumann [2], Schroedinger [4], and so on. Furthermore, with the recent development of quantum information theory [18] and experimental techniques to examine quantum mechanics, the typicality has been attracting much more attentions. However, "genuine thermodynamic variables" such as entropy and temperature cannot be obtained in these studies. Without the genuine thermodynamic variables, a new formulation of statistical mechanics cannot be made.

In this thesis, I establish the formulation of statistical mechanics based on a pure quantum state [19–21]. In this formulation, I fully utilize the typicality in quantum systems. I generally call the pure quantum states which are regarded as the equilibrium states "thermal pure quantum (TPQ) states." I extend the typicality so that it is applicable to the genuine thermodynamic variables such as entropy and temperature. Thus, not only the mechanical variables such as energy and magnetization, but also the genuine thermodynamic variables can be obtained using a *single* realization of the TPQ state. I introduce many kinds of the TPQ states which include a "(grand)canonical TPQ state," which is specified by intensive variables such as temperature, and a "microcanonical TPQ state," which is easy to generate. Owing to these results, I establish the new formulation of statistical mechanics based on the TPQ states. These findings do not only deepen our understanding of the foundation of statistical mechanics, but also enable efficient numerical calculations. I apply the TPQ formulation to the the spin-1/2 kagome Heisenberg antiferromagnet and find out a new behavior of the specific heat.

The equilibrium state is represented by mixed quantum states in the conventional formulation using the ensembles. By contrast, the TPQ states are pure quantum states, and thus, they are completely different from the conventional equilibrium states in microscopic sense. In fact, we can detect the difference between them by measuring quantum entanglement, which quantifies quantum-mechanical correlations. The entanglement of the TPQ states are large, and especially at high temperature, almost maximum. On the other hand, the entanglement of the conventional equilibrium states are small and vanishing to zero as temperature increases. Namely, the TPQ formulation employs the completely different microstates from the conventional ones. Nevertheless, both microstates are regarded as the same equilibrium state as far as we look at the macroscopic quantities.

The contents of this thesis are as follows: In Chap. 2, I briefly survey the discussions of the foundation of statistical mechanics. Among them, the typicality in quantum system gives the strongest predictions. In Chap. 3, I define the TPQ state and introduce the cTPQ state. In Chap. 4, I introduce another TPQ state, the mTPQ state. The mTPQ state puts more weight on practical applications than the cTPQ state. In Chap. 5, the physical properties of the TPQ state are discussed further. The TPQ states are applicable to the linear response theory. Here, the TPQ formulation works just like the ensemble formulation. However, fluctuation of the TPQ states and that of the states in the ensemble formulation are represented in different forms. Moreover, when we observe the quantum entanglement, these states are completely different. In Chap. 6, I give transformation formulas among the TPQ states. These transformations enable us to perform the efficient calculations using the TPQ states. In Chap. 7, I examine the TPQ formulation as a numerical method. The TPQ formulation is applied to the spin-1/2 kagome Heisenberg antiferromagnet (KHA), which is one of the popular frustrated quantum models.

References

1. L. Boltzmann, Wiener Berichte **76**, 373 (1877), Reprinted in F. Hasenohrl ed. J. A. Barth **2**, 164 (1909)
2. J. von Neumann, Zeitschrift fuer Physik **57**, 30 (1929) (In German), J. von Neumann, Eur. Phys. J. H **35**, 173 (2010) (English Translation)
3. A.I. Khinchin, *Mathematical Foundations of Statistical Mechanics Chap. 3* (Dover, New York, 1949), pp. 44–47
4. E. Schroedinger, *Statistical Thermodynamics* (Cambridge University Press, Cambridge, England, 1952)
5. S. Goldstein, J.L. Lebowitz, R. Tumulka, N. Zanghi, Eur. Phys. J. **35**, 201 (2010)
6. S. Goldstein, J.L. Lebowitz, C. Mastrodonato, R. Tumulka, N. Zanghi, Proc. R. Soc. A **466**(2123), 3203 (2010)
7. P. Bocchieri, A. Loinger, Phys. Rev. **114**, 948 (1959)
8. S. Lloyd, Black Holes, *Demons, and the Loss of Coherence. Chap. 3* (Thesis, Rockefeller Univ. 1988), pp. 79–93
9. H. Tasaki, Phys. Rev. Lett. **80**, 1373 (1998)
10. J. Gemmer, G. Mahler, Euro. Phys. J. B **31**, 249 (2003)
11. J. Gemmer, M. Michel, G. Mahler, *Quantum Thermodynamics*. Lecture Notes in Physics, vol. 657 (Springer, Berlin, 2004)
12. S. Popescu, A.J. Short, A. Winter, Nat. Phys. **2**, 754 (2006)
13. S. Goldstein et al., Phys. Rev. Lett. **96**, 050403 (2006)
14. A. Sugita, RIMS Kokyuroku (Kyoto) **1507**, 147 (2006). (in Japanese)
15. A. Sugita, Nonlinear Phenom. Complex Syst. **10**, 192 (2007)
16. P. Reimann, Phys. Rev. Lett. **99**, 160404 (2007)
17. P. Reimann, J. Stat. Phys. **132**, 921 (2008)
18. M.A. Nielsen, I.L. Chuang, *Quantum Computation and Quantum Information* (Cambridge University Press, Cambridge, 2000)
19. S. Sugiura, A. Shimizu, Phys. Rev. Lett. **108**, 240401 (2012)
20. S. Sugiura, A. Shimizu, Phys. Rev. Lett. **111**, 010401 (2013)
21. S. Sugiura, A. Shimizu, Kinki University Series on Quantum Computing. **9**, 245 (2014), arXiv:1312.5145

Chapter 2
Typicality and Ergodicity

2.1 Typicality and Ergodicity in Classical Statistical Mechanics

The PEPP is the postulate that all the possible microscopic states appear in the same probability. Since all the probabilities for the states are severely restricted by this postulate, it is counter-intuitive and thus many people have tried to justify it microscopically. In many textbooks, ergodicity of the state is often referred for the justification of PEPP. However, it is not physically realistic and cannot be the justification, although ergodic theorem itself has brought fruitful mathematics. Instead of the ergodicity, I introduce the typicality for the interpretation of PEPP. In this section, I review these points in a classical setup.

2.1.1 Ergodicity and Its Problems

When we employ a microcanonical ensemble in statistical mechanics, an equilibrium value is obtained by microcanonical ensemble average,

$$\langle A\rangle_{\mathrm{mc}} \equiv \frac{\int_{E-\Delta E<E(\{p_i\}_i,\{q_i\}_i)\le E} A\left(\{p_i\}_i,\{q_i\}_i\right)\mathrm{d}\Gamma}{W} \tag{2.1}$$

where $(\{p_i\}_i, \{q_i\}_i)$ is a set of positions p_i and momentums q_i for all particles (i is an index of particles), $A\left(\{p_i\}_i, \{q_i\}_i\right)$ is a value of a physical variable A for $(\{p_i\}_i, \{q_i\}_i)$, E is energy, $(E-\Delta E, E]$ is an energy shell, $\mathrm{d}\Gamma \equiv \Pi_i \mathrm{d}p_i \mathrm{d}q_i$, and W is the number of states

$$W \equiv \int_{E-\Delta E<E(\{p_i\}_i,\{q_i\}_i)\le E} \mathrm{d}\Gamma. \tag{2.2}$$

S. Sugiura, *Formulation of Statistical Mechanics Based on Thermal Pure Quantum States*, Springer Theses, DOI 10.1007/978-981-10-1506-9_2

The averaged value over time is

$$\langle A \rangle_{\text{time}} \equiv \lim_{T\to\infty} \frac{1}{T} \int_0^T A\left(\{p_i(t)\}_i, \{q_i(t)\}_i\right) \mathrm{d}t \tag{2.3}$$

where $(\{p_i(t)\}_i, \{q_i(t)\}_i)$ is a set of positions and momentums at time t. Ergodicity is the property that the microcanonical ensemble average value of every physical variable A is equal to the time average value of it, i.e., for $^\forall A$,

$$\langle A \rangle_{\text{mc}} = \langle A \rangle_{\text{time}}. \tag{2.4}$$

In Newtonian mechanics, the ergodicity is proved for some models. However, even if Eq. (2.4) holds, the time for the state $(\{p_i(t)\}_i, \{q_i(t)\}_i)$ to evolve over all the possible states T_{evolve} is usually too long to measure. T_{evolve} easily exceeds the lifetime of the universe. Namely, we cannot wait for T_{evolve} coming, even if the ergodicity holds. On the other hand, we do not need to measure the physical quantities for a long time $\sim T_{\text{evolve}}$ to know the equilibrium values in reality. For example, when we measure the temperature of a glass of water for a minute, we can obtain the equilibrium value of it. Therefore, the ergodicity will not be the reason why the ensemble average gives the correct equilibrium value. In addition to this, T_{evolve} increases exponentially as the size of the system increases. In contrast to this, statistical mechanics is an asymptotic theory whose prediction gets closer to thermodynamics as the system size increases. Hence, the ergodicity seems to have nothing to do with the foundation of statistical mechanics.

2.1.2 *Typicality in Weak Sense*

In Sect. 2.1.1, we saw that ergodicity cannot be used for the basis of statistical mechanics. Alternatively, we can easily prove that "weak typicality" among states can be substituted for the PEPP.

The weak typicality is stated as follows: Let a macroscopic physical variable A be extensive, that is, when a system gets x-times larger, a value of A also gets x-times larger. For example, energy, magnetization, number of particle are the extensive variables. I say that the weak typicality for A holds if $^\forall \varepsilon > 0$, $^\exists$system size V, such that

$$\frac{\sqrt{\Delta A^2}}{V} < \varepsilon \tag{2.5}$$

where

$$\Delta A^2 \equiv \frac{\int \{A(\{p_i(t)\}_i, \{q_i(t)\}_i) - \langle A \rangle_{\text{ens}}\}^2 \,\mathrm{d}\Gamma}{W}. \tag{2.6}$$

Since the number of particles increases linearly as V increases, $\sqrt{\Delta A^2}$ also increases as V does. However, the weak typicality postulates that $\sqrt{\Delta A^2} = o(V)$. It means that if I prepare a state for sufficiently large V randomly among all possible states,

$$a(\{p_i(t)\}_i, \{q_i(t)\}_i) \equiv \frac{A(\{p_i(t)\}_i, \{q_i(t)\}_i)}{V} \tag{2.7}$$

rarely deviates from its ensemble average. Therefore, I do not need to assume the PEPP to obtain the thermodynamic predictions from Newtonian mechanics. Since the PEPP restricts all the realizing probability of the possible states, it is a stronger assumption than the weak typicality. If I only assume the weak typicality instead of PEPP, $a(\{p_i(t)\}_i, \{q_i(t)\}_i)$ takes the value which is equal to the ensemble average within a negligible error.

Inversely, I can also show that the weak typicality is proved from PEPP. I assume that PEPP, the Boltzmann's entropy formula, and the convexity of entropy, i.e.,

$$\frac{\partial^2 S(E, A, V)}{\partial A^2} \leq 0, \tag{2.8}$$

where $S(E, A)$ is entropy and I omit variables except for energy E, some physical quantity A, and the volume V. By the definition of the Boltzmann's formula, I get

$$W(E, A, V) = \exp[S(E, A, V)] \tag{2.9}$$

$$= \exp[V\{s(u, a; V)\}] \tag{2.10}$$

where $s(u, a; V) \equiv S(E, A, V)/V$, $u \equiv E/V$, and $a \equiv A/V$. $s(E, A; V)$ take the maximum at a^* where

$$\left.\frac{\partial s(u, a; V)}{\partial a}\right|_{a=a^*} = 0. \tag{2.11}$$

Expanding the rhs of Eq. (2.10) around a^*, I get

$$W(E, A, V) = \exp[V\{s(u, a^*; V) + s''(u, a^*; V)(a - a^{*2}) + O\left((a - a^{*3})\right)\}] \tag{2.12}$$

where

$$s''(u, a^*; V) \equiv \left.\frac{\partial^2 s(u, a; V)}{\partial a^2}\right|_{a=a^*} < 0 \tag{2.13}$$

and $s''(u, a^*; V)$ is $\Theta(V^0)$. Therefore, $W(E, A, V)$ behaves like a Gaussian distribution with a peak at

$$a = a^* \quad (\Leftrightarrow A = Va^*) \tag{2.14}$$

and the width

$$\Delta a^2 = \Theta(\frac{1}{V}) \quad (\Leftrightarrow \Delta A^2 = \Theta(V)). \tag{2.15}$$

When one specifies the value of energy but does not specify that of a, a may take various values. However, since $\lim_{V\to\infty} \Delta a^2 = 0$ in Eq. (2.15), not only the ensemble average takes a^* but also the values of a of almost all the possible states take a^* with errors vanishing to zero with increasing V. This is the proof of the weak typicality for a derived from the above three assumptions, PEPP, the Boltzmann's entropy formula, and the convexity of entropy. Since the Boltzmann's entropy formula and the convexity of entropy are always assumed in a usual situation, the PEPP automatically implies the weak typicality.

In summary, the weak typicality is sufficient for obtaining thermodynamic predictions. Although the PEPP is the sufficient condition for the weak typicality, it is not the necessary condition for it. Therefore, one may expect that the weak typicality is more fundamental than the PEPP in statistical mechanics. However, this discussion of the weak typicality focuses only on the extensive variables. We cannot apply the above discussion to other variables; for example, fluctuations, one-particle observables like its position, and inensive variables are excluded. I do not know whether this discussion is extended to these variables. Thus, the justification of statistical mechanics using the weak typicality in this form is not completed.

2.2 Typicality and Ergodicity in Quantum Statistical Mechanics

In Sect. 2.1, I have explained that neither ergodicity nor PEPP is necessary to obtain thermodynamic predictions of some extensive variables in classical system, but the weak typicality is sufficient. For quantum systems, I will show that ergodicity generally breaks down [1–5], but typicality is proved in a stronger sense than what I have explained in classical statistical mechanics [6–11]. I call it "strong typicality." The strong typicality is stated as follows: First, the expectation value $\langle|\psi|\hat{A}\psi\rangle$ for any "mechanical variables" (defined in this section) is close to the microcanonical ensemble average under a natural probability measure of the coefficients $\{c_n\}_n$. Second, the difference between $\langle|\psi|\hat{A}\psi\rangle$ and the ensemble average is of the order $\exp[\Theta(-V)]$, which immediately vanishes to zero as V increases. It will turn out that the strong typicality can substitute for the PEPP as a principle of statistical mechanics.

2.2.1 *Ergodicity in Quantum Mechanics*

Let me consider an arbitrary quantum pure state $|\psi_{\mathrm{erg}}\rangle$ in a energy shell $(E-\delta E, E]$. It can be written in a general form,

$$|\psi_{\mathrm{erg}}\rangle \equiv \sum_{n\in\mathscr{S}} c_n|n\rangle. \qquad \mathscr{S} = \{n|u_n \in (E-\delta E, E]\} \tag{2.16}$$

where $\{c_n\}_n$ is a set of coefficients and $|n\rangle$ is an energy eigenstate, $\hat{H}|n\rangle = u_n|n\rangle$. The time evolution of $|\psi\rangle$ is

$$|\psi_{\text{erg}}(t)\rangle \equiv \sum_n c_n e^{-\frac{i}{\hbar}u_n t}|n\rangle, \tag{2.17}$$

and the expectation value of $|\psi(t)\rangle$ for an operator $\hat{A}$ is

$$\langle\psi_{\text{erg}}(t)|\hat{A}|\psi_{\text{erg}}(t)\rangle = \sum_{n,m} c_n^* c_m e^{\frac{i}{\hbar}(u_n - u_m)t}\langle n|\hat{A}|m\rangle. \tag{2.18}$$

Then, the time average of the expectation value is

$$\begin{aligned}\langle\hat{A}\rangle_{\text{time}} &\equiv \lim_{T\to\infty}\frac{1}{T}\int_0^T \langle\psi_{\text{erg}}(t)|\hat{A}|\psi_{\text{erg}}(t)\rangle \\ &= \sum_{n,m} c_n^* c_m \langle n|\hat{A}|m\rangle \lim_{T\to\infty}\frac{1}{T}\int_0^T e^{\frac{i}{\hbar}(u_n-u_m)t} && (2.19)\\ &= \sum_{n,m} c_n^* c_m \langle n|\hat{A}|m\rangle \delta_{n,m} && (2.20)\\ &= \sum_n |c_n|^2 \langle n|\hat{A}|n\rangle && (2.21)\end{aligned}$$

On the other hand, the microcanonical ensemble average is

$$\langle\hat{A}\rangle_{\text{mc}} \equiv \frac{1}{d}\sum_n \langle n|\hat{A}|n\rangle. \tag{2.22}$$

where $d = \dim\mathscr{S}$. Therefore,

$$\langle\hat{A}\rangle_{\text{time}} \neq \langle\hat{A}\rangle_{\text{mc}}, \tag{2.23}$$

holds for *all* observable $\hat{A}$, if and only if

$$|c_n|^2 = \frac{1}{d} \tag{2.24}$$

for $^\forall n$. Except for the case of Eq. (2.24), the ergodicity does not hold in quantum statistical mechanics, As I have emphasized, ergodicity is too strict and unnecessary statement. The failure of the ergodicity is because the ergodicity tries to prove Eq. (2.23) for all observables, which include arbitrary N-body correlations. For the foundation of statistical mechanics, therefore, we have to think seriously what kind of operators we look at. This leads to the understanding of the strong typicality.

2.2.2 Strong Typicality in Quantum System

We examine the typicality in quantum systems. The discussion in Sect. 2.2.2 mostly follows Refs. [8]. We consider a state $|\psi_E\rangle$ in an energy shell;

$$|\psi_E\rangle \equiv \sum_{n\in\mathscr{S}} c_n|n\rangle. \qquad \mathscr{S} = \{n|u_n \in (E-\delta E, E]\} \tag{2.25}$$

Here, $\{c_n\}_n$ is a set of coefficients labeled by n, $|n\rangle$ and u_n are energy eigenstates and corresponding energy eigenvalues, respectively. Since the question "Is it typical?" depends on a measure. We introduce the probability measure $\mathrm{P}(\{c_n\}_n)$ into $\{c_n\}_i$. Although $|\psi_E\rangle$ looks the same as $|\psi_{\text{erg}}\rangle$, $\{c_n\}_n$ in $|\psi_E\rangle$ is random variables in this case. In order to make the measure be natural for quantum systems, we impose two conditions to $\mathrm{P}(\{c_n\}_n)$: (1) Invariance—The measure should be invariant under arbitrary unitary transformations. (2) Normalization—$\{c_n\}_n$ should be normalized; $\sum_i |c_n|^2 = 1$. Therefore, $\mathrm{P}(\{c_n\}_n)$ should be

$$\mathrm{P}(\{c_n\}_n) = \frac{\Gamma(d-1/2)}{2\pi^{d-1}}\,\delta\!\left(\sum_n |c_n|^2 - 1\right), \tag{2.26}$$

where $d = \dim\mathscr{S}$.

Under this measure [13], the variance of $\langle\psi_E|\hat{A}|\psi_E\rangle - \langle\hat{A}\rangle_{\text{MC}}$ is calculated as

$$\overline{\left(\langle\psi_E|\hat{A}|\psi_E\rangle - \langle\hat{A}\rangle_{\text{MC}}\right)^2} = \frac{\langle\Delta\hat{A}^2\rangle_{\text{MC}} + \langle\hat{A}\rangle_{\text{MC}}^2}{d+1}, \tag{2.27}$$

where $\overline{\cdots}$ denotes the random average of $\cdots$, and $\langle\Delta\hat{A}^2\rangle_{\text{MC}} \equiv \langle(\hat{A}-\langle\hat{A}\rangle_{\text{MC}})^2\rangle_{\text{MC}}$. In Eq. (2.27), the denominator of the rhs is $\exp[\Theta(V)]$, and the numerator is the order of polynomial of V for any mechanical variable (defined in Sect. 2.2.3). Hence,

$$\text{the rhs of Eq. (2.27)} = \exp[-\Theta(V)]. \tag{2.28}$$

This result means that almost all $|\psi_E\rangle$'s give correct results which are close to the microcanonical ensemble average. In this sense, the typicality holds. Moreover, the error shown in Eq. (2.27) is exponentially small as V increases. This error is much smaller than $\Theta(\frac{1}{V})$, which is the error of the weak typicality for classical systems in Sect. 2.1.2. Hence, I call this typicality a "strong typicality." The strong typicality is the most important property for statistical mechanics based on pure quantum states.

I note that the measure given in Eq. (2.26) is simple and natural but not necessary. I expect that the typicality will hold for other reasonable probability measures of $\{c_n\}_n$ I also note a possible interpretation of $|\psi_E\rangle$. In the ensemble formulation of quantum statistical mechanics, equilibrium state corresponding to microcanonical ensemble is

$$\hat{\rho}_{\mathrm{MC}} \equiv \frac{1}{d} \sum_{n \in \mathscr{S}} |n\rangle\langle n|. \tag{2.29}$$

Then, I can show that

$$\hat{\rho}_{\mathrm{MC}} = \overline{|\psi_E\rangle\langle\psi_E|}. \tag{2.30}$$

Hence, $|\psi_E\rangle$ can be interpreted as a single realization of a pure quantum state from $\hat{\rho}_{\mathrm{MC}}$ of the form of Eq. (2.30).

2.2.3 *Mechanical Variables and Genuine Thermodynamic Variables*

In a macroscopic system, we do not measure all of the possible observable, but measure several macroscopic observables such as magnetization and correlation functions. I call these macroscopic observables as "mechanical variables (MVs)" and define them in a rigorous way [12].

Def.) Mechanical Variables
A mechanical variable is a quantum-mechanical observable which is low-degree polynomials of local operators and satisfies the condition that

$$\langle \hat{A}^2 \rangle^{\mathrm{ens}}_{\beta,V} \leq K(\beta) V^{2m} \text{for all } \beta, \mathrm{V}. \tag{2.31}$$

Here, a function $K(\beta)$ and a constant m are positive and independent of $\hat{A}$ and V. I make $\hat{A}$ dimensionless by dividing an appropriate unit. Since I have fixed m, the number of independent MVs N_{MV} is $\Theta(V^m)$.

Returning to the main result, we can rewrite Eq. (2.27) by using Markov's inequality as

$$\mathrm{P}\left(\left|\langle\psi_E|\hat{A}|\psi_E\rangle - \langle\hat{A}\rangle_{\mathrm{MC}}\right| \geq \varepsilon\right) \leq \frac{1}{\varepsilon^2} \frac{\langle\Delta\hat{A}^2\rangle_{\mathrm{MC}} + \langle\hat{A}\rangle^2_{\mathrm{MC}}}{d+1}, \tag{2.32}$$

where $\mathrm{P}(\cdots)$ denotes the probability of event $\cdots$. Therefore, the probability that all expectation values of the MVs for $|\psi_E\rangle$ do not deviate from the corresponding ensemble averages is estimated as follows:

$$\mathrm{P}\left(\sum_{\hat{A}\in\mathrm{MV}} \left|\langle\psi_E|\hat{A}|\psi_E\rangle - \langle\hat{A}\rangle_{\mathrm{MC}}\right| \geq \varepsilon\right) \leq \frac{N^2_{\mathrm{MV}}}{\varepsilon^2} \frac{1}{d+1} \max_{\hat{A}\in\mathrm{MV}} \left(\langle\Delta\hat{A}^2\rangle_{\mathrm{MC}} + \langle\hat{A}\rangle^2_{\mathrm{MC}}\right). \tag{2.33}$$

Since N_{MV} is $\Theta(V^m)$, the rhs is still $\exp[\Theta(-V)]$. Hence, a single realization of $|\psi_E\rangle$ for sufficiently large V correctly gives all equilibrium values of the MVs simultaneously.

The number of mechanical variables N_{MV} is exponentially smaller than d, which is the dimension of the Hilbert space in the energy shell. Thus, even when we measure all the MVs, we cannot identify a pure quantum state but almost all $|\psi_E\rangle$ return the same value for all the MVs.

However, entropy and temperature are not included in this discussion, because they cannot be represented as a MV. If one stick to represent these variables as quantum mechanical observables, they become N-body operators in general, while low-degree polynomials of local operators are the only MVs. In contrast to quantum mechanics, any equilibrium values for macroscopic variables including entropy and temperature are obtained from any of the thermodynamic functions in thermodynamics. I call these macroscopic variables except for the MVs "genuine thermodynamic variables (GTVs)" [12].

Def.) Genuine Thermodynamic Variables
Genuine thermodynamic variables are observables which are not MVs but can be derived from entropy.

For example, (inverse) temperature is the derivative of entropy with respect to energy. Free energy is obtained by Legendre transformation of entropy. Hence, they are GTVs.

2.2.4 *Canonical Typicality*

Using the results in Sect. 2.2.2, it is revealed that the subsystem of the $|\psi_E\rangle$ is almost identical to the Gibbs state in that subsystem [6, 7, 14–18]. Let the total system be a composite of system (S) and environment (E), and the interaction between S and E be negligible. The size of the system is v, and the size of the environment is V. Here, we consider the case $v \ll V$. We assume the Hamiltonian

$$\hat{H} \equiv \hat{H}_S + \hat{H}_E, \tag{2.34}$$

where the support of $\hat{H}_S$ is on S and that of $\hat{H}_E$ is on E. We prepare $|\psi_E\rangle$ of the total system. The reduced density matrix of $|\psi_E\rangle$ in S is

$$\hat{\rho}_S \equiv \text{Tr}_B[|\psi_E\rangle\langle\psi_E|]. \tag{2.35}$$

Next, let $\hat{A}_S$ be an arbitrary observable in S. N_S, which is the number of $\hat{A}_S$, scales as $O(\nu^v)$ while the total dimension of Hilbelt space d in Eq. (2.33) scales as $O(\nu^V)$. Here, ν is a constant, e.g., $\nu = 2$ for the spin-1/2 model and $\nu = 4$ for the Hubbard model. Hence, using Inequality (2.33), we get

$$\mathrm{P}\left(\sum_{\hat{A}_S}\left|\langle\psi_E|\hat{A}_S|\psi_E\rangle - \langle\hat{A}_S\rangle_{\mathrm{MC}}\right| \geq \varepsilon\right) \leq \frac{N_S^2}{\varepsilon^2}\frac{1}{d+1}\max_{\hat{A}_S}\left(\langle\Delta\hat{A}_S^2\rangle_{\mathrm{MC}} + \langle\hat{A}_S\rangle_{\mathrm{MC}}^2\right). \tag{2.36}$$

Since the rhs is $\exp[\Theta(-V)]$, we show that $|\psi_E\rangle$ gives equilibrium value for all the observable in S. I note that $\hat{A}_S$ includes ν-body correlations, which is a microscopic quantities for the system S. Since $\nu \ll V$, the ν-body correlations are low-degree polynomial for the total system. Therefore, $|\psi_E\rangle$, which is the pure state of the total system can give *all* equilibrium value for observables in S.

When two quantum states give the same expectation value for all the possible observable in the subsystem S, these state are identical in S. Namely, $\hat{\rho}_S \simeq e^{-\beta\hat{H}_S}/\mathrm{Tr}[e^{-\beta\hat{H}_S}]$ where β is temperature such that

$$\langle\hat{H}_E\rangle_{\beta,V}^{\mathrm{can}} = E. \tag{2.37}$$

The expectation value of any observable $\hat{A}$ in S gives

$$\langle\psi_E|\hat{A}|\psi_E\rangle = \langle\hat{A}\rangle_{\beta,V}^{\mathrm{can}} \tag{2.38}$$

with the exponentially small error. This is so-called "canonical typicality" [7].

2.3 Summary and Discussions of this Chapter

In this section, we saw that the ergodicity is physically nonsense for the justification of the PEPP. Moreover, it breaks down in quantum system. Thus, we saw the typicality among states. Even in classical system, we can show the weak typicality, which is enough to obtain thermodynamic predictions for some extensive variables. However, it may not be sufficient for the foundation of statistical mechanics. In quantum system, the strong typicality was shown. Owning to the exponentially large Hilbert space, all equilibrium values of the MVs, which include even a position of a particle and higher order fluctuation, can be predicted from a single pure state in the strong typicality. The strong typicality can be another principle of statistical mechanics by substituting the PEPP. It will be also expected that the exponentially small error enable us to apply such a pure state to practical applications.

However, the GTVs such as entropy are excluded from the MVs. From any of the thermodynamic functions, all the thermodynamic predictions are derived. Since the genuine thermodynamic variables are related closely to the number of states through Boltzmann entropy formula, it may seem to be difficult to obtain them from a single pure quantum state. In addition to this, we do not know whether there exist pure quantum states which are specified by intensive variables such as temperature and chemical potential. For practical applications, it is also preferred that such pure quantum states can be generated easily, but we do not know how. I will answer these questions positively from now on. Here, I summarize the important points in this chapter:

Lesson:
As far as we look at MVs, we can neither distinguish different realizations of $|\psi_E\rangle$'s, nor do $|\psi_E\rangle$ from $\hat{\rho}_{\mathrm{MC}}$.
Finding Of Previous Works:
The probabilistic error of $|\psi_E\rangle$ is $\exp[\Theta(-V)]$, which is much smaller than that of the weak typicality, $O(\frac{1}{\sqrt{V}})$.
Questions:
Can we obtain the genuine thermodynamic variables from a single realization of a pure quantum state?
By specifying the intensive variables, can we generate pure states which are regarded as equilibrium states ?
How to construct these states?

References

1. J. von Neumann, Zeitschrift fuer Physik **57**, 30 (1929) (in German); J. von Neumann, Eur. Phys. J. H **35**, 173 (2010) (English Translation)
2. S. Goldstein, J.L. Lebowitz, R. Tumulka, N. Zanghi, Eur. Phys. J. H **35**, 201 (2010)
3. S. Goldstein, J.L. Lebowitz, C. Mastrodonato, R. Tumulka, N. Zanghi, Proc. R. Soc. A **466**(2123), 3203 (2010)
4. P. Reimann, Phys. Rev. Lett. **101**, 190403 (2008)
5. Z.X. Gong, L.-M. Duan, arXiv:1109.4696
6. S. Popescu, A.J. Short, A. Winter, Nat. Phys. **2**, 754 (2006)
7. S. Goldstein et al., Phys. Rev. Lett. **96**, 050403 (2006)
8. A. Sugita, RIMS Kokyuroku (Kyoto) **1507**, 147 (2006). (in Japanese)
9. A. Sugita, Nonlinear Phenom. Complex Syst. **10**, 192 (2007)
10. P. Reimann, Phys. Rev. Lett. **99**, 160404 (2007)
11. P. Reimann, J. Stat. Phys. **132**, 921 (2008)
12. S. Sugiura, A. Shimizu, Phys. Rev. Lett. **108**, 240401 (2012)
13. N. Ullah, Nucl. Phys. **58**, 65 (1964)
14. E. Lubkin, J. Math. Phys. **19**, 1028 (1978)
15. S. Lloyd, H. Pagels, Ann. Phys. **188**, 186 (1988)
16. D.N. Page, Phys. Rev. Lett. **71**, 1291 (1993)
17. S. Sen, Phys. Rev. Lett. **77**, 1 (1996)
18. H.-J. Sommers, K. Zyczkowski, J. Phys. A: Math. Gen. **37**, 8457 (2004)

Chapter 3
Canonical Thermal Pure Quantum State

As I have briefly surveyed in Chap. 2, there is the possibility that a pure quantum state can represent the equilibrium state. In this section, I define the pure state which can be regarded as the equilibrium state. I call it a thermal pure quantum (TPQ) state [1]. I also introduce other TPQ states corresponding to other ensembles [2]. I show that the equilibrium values of both the mechanical variables (MVs) and the genuine thermodynamic variables (GTVs) are obtained from a *single* realization of the TPQ state. Using these findings, I establish statistical mechanics based on the TPQ states [2].

3.1 Definition of Thermal Pure Quantum State

I define the TPQ state as follows [1]:

Def.) Thermal Pure Quantum State
Consider a state $|\psi\rangle$ which has random variables, For $^\forall\epsilon > 0$, if $^\exists\eta_V(\epsilon)$ such that $^\forall\hat{A} \in$ MVs,

$$\mathrm{P}\left(\left|\frac{\langle\psi|\hat{A}|\psi\rangle}{\langle\psi|\psi\rangle} - \langle\hat{A}\rangle_{\mathrm{ens}}\right| \geq \epsilon\right) \leq \eta_V(\epsilon) \tag{3.1}$$

and $\eta_V(\epsilon) \overset{V\to\infty}{\to} 0$, I call $|\psi\rangle$ a TPQ state.

Here, $\langle\cdots\rangle_{\mathrm{ens}}$ is the ensemble average. For example, $|\psi_E\rangle$ in the Sect. 2.2 is a TPQ state, by definition. However, it is not the only TPQ state corresponds to the microcanonical ensemble, but various kinds of pure quantum states can be the one.

S. Sugiura, *Formulation of Statistical Mechanics Based on Thermal Pure Quantum States*, Springer Theses, DOI 10.1007/978-981-10-1506-9_3

Furthermore, since I do not restrict the ensembles to the microcanonical one in Eq. (3.1), there can be TPQ states correspond to the canonical and other ensembles. These TPQ states are the main focus in this chapter.

3.2 Introduction of Canonical Thermal Pure Quantum State

In this subsection, I introduce a TPQ state corresponds to the canonical ensemble. I call it a canonical TPQ (cTPQ) state. Firstly, I consider

$$|i;\beta,V\rangle = \exp\left[-\frac{\beta\hat{H}}{2}\right]|i\rangle, \tag{3.2}$$

where $\{|i\rangle\}_i$ is a set of arbitrary orthonormal basis. The exponential function $\exp\left[-\frac{\beta\hat{H}}{2}\right]$ works as an energy cutoff. However, when I choose $|i\rangle$ randomly among $\{|i\rangle\}_i$, this state may not be regarded as the TPQ state. To show this, I use these two results.

$$\frac{\overline{\langle i;\beta,V|\hat{A}|i;\beta,V\rangle}^i}{\overline{\langle i;\beta,V|i;\beta,V\rangle}^i} = \langle\hat{A}\rangle^{\mathrm{ens}}_{\beta,V}, \tag{3.3}$$

$$\overline{\left(\frac{\langle i;\beta,V|\hat{A}|i;\beta,V\rangle}{\langle i;\beta,V|i;\beta,V\rangle} - \langle\hat{A}\rangle^{\mathrm{ens}}_{\beta,V},\right)^2}^i \geq \overline{\left(\frac{\langle i;\beta,V|\hat{A}|i;\beta,V\rangle}{Z(\beta,V)} - \langle\hat{A}\rangle^{\mathrm{ens}}_{\beta,V},\right)^2}^i \tag{3.4}$$

$$\leq \langle(\Delta\hat{A})^2\rangle^{\mathrm{ens}}_{\beta,V} \tag{3.5}$$

where $\overline{\cdots}^i$ denotes the average over all $\{|i\rangle\}_i$, $\langle\hat{A}\rangle^{\mathrm{ens}}_{\beta,V} \equiv \frac{\mathrm{Tr}\left[\exp(-\beta\hat{H})\hat{A}\right]}{Z(\beta,V)}$, $Z(\beta,V) \equiv \mathrm{Tr}[\exp(-\beta\hat{H})]$, and $\langle(\Delta\hat{A})^2\rangle^{\mathrm{ens}}_{\beta,V} \equiv \langle(\hat{A} - \langle A\rangle^{\mathrm{ens}}_{\beta,V})^2\rangle^{\mathrm{ens}}_{\beta,V}$.

I evaluate the lhs of inequality (3.4). Firstly, the equality in inequality (3.5) is attained, for example, if $\{|i\rangle\}_i$ is a set of energy eingenstates and $\hat{A}$ is Hamiltonian $\hat{H}$. When this equality is attained, I get

$$\overline{\left(\frac{\langle i;\beta,V|\hat{A}|i;\beta,V\rangle}{\langle i;\beta,V|i;\beta,V\rangle} - \langle\hat{A}\rangle^{\mathrm{ens}}_{\beta,V},\right)^2}^i \geq \langle(\Delta\hat{A})^2\rangle^{\mathrm{ens}}_{\beta,V}. \tag{3.6}$$

Therefore, what I need to evaluate is $\langle(\Delta\hat{A})^2\rangle^{\mathrm{ens}}_{\beta,V}$. When the system is consistent with thermodynamics [3], we can usually expect Eq. (2.15), that is,

$$\frac{\sqrt{\langle(\Delta\hat{A})^2\rangle^{\text{ens}}_{\beta,V}}}{\langle\hat{A}\rangle^{\text{ens}}_{\beta,V}} = O\left(\frac{1}{\sqrt{V}}\right). \tag{3.7}$$

Using Eq. (3.7) and $\langle\hat{A}\rangle^{\text{ens}}_{\beta,V} = O(V^m)$, the lhs of inequality (3.4) can be larger than $\Theta(V^{2m-1})$. The condition (3.1) is violated in this case. Hence, $|i;\beta,V\rangle$ is not always the TPQ state. The failure of $|i;\beta,V\rangle$ is because I employed some specific basis $\{|i\rangle\}$. For example, when I employ $\{|i\rangle\}$ as a set of product states and $[\hat{H},\hat{S}_z]=0$, $|i;\beta,V\rangle$ conserves S_z of the initial state $|i\rangle$. Thus, $|i;\beta,V\rangle$ can have an initial state dependence. In order to avoid this dependence, I should use a set of random vector as the basis.

Learning from this lesson, I consider a state $|\beta,V\rangle$ which is the superposition of all $|i;\beta,V\rangle$'s with random coefficients z_i's.

$$|\beta,V\rangle \equiv \sum_i z_i \exp\left[-\frac{\beta\hat{H}}{2}\right]|i\rangle. \tag{3.8}$$

Here, $z_i \equiv \frac{(x_i+iy_i)}{\sqrt{2}}$ is a random variable, where x_i and y_i are real random variables, each obeying the unit normal distribution, and $\{|i\rangle\}_i$ is an arbitrary orthonormal basis set spanning Hilbert space. Since the probability measure of $\{z_i\}_i$ is invariant under any unitary transformation, the dependence on the basis $|i\rangle$ vanishes owing to the introduction of $\{z_i\}_i$. I will show in Sect. 3.3 that $|\beta,V\rangle$ satisfies the condition of the TPQ state. I call this the canonical TPQ state.

In the conventional statistical mechanics using the ensemble formulation, the equilibrium state specified by temperature is represented by a Gibbs state,

$$\hat{\rho}_\beta \equiv \frac{\exp[-\beta\hat{H}]}{Z(\beta,V)}. \tag{3.9}$$

In Sect. 3.3, I will show the following two results. First, using the cTPQ state $|\beta,V\rangle$, the genuine thermodynamic variables are obtained from the length of the vector $|\beta,V\rangle$. Second, MVs are obtained from its expectation values. In this sense, the cTPQ state is the counterpart of the Gibbs state in the formulation of statistical mechanics based on the TPQ states. Although I do not assume the PEPP, which is one of the basic principles in the ensemble formulation, I will reveal in Sect. 3.4 that the random variables $\{z_i\}_i$ substitute for the PEPP in the TPQ formulation.

I notice that the TPQ formulation is applicable even if the dimension of Hilbert space is infinite [4]. Since $|\beta,V\rangle$ is the superposition of infinite number of states in such a case, one might think its norm would diverge and $|\beta,V\rangle$ would be ill-defined. However, the exponential function $\exp\left[-\frac{\beta\hat{H}}{2}\right]$ cuts off high energy states and the norm of $|\beta,V\rangle$ becomes finite.

3.3 Main Results

In this section, I see some important results of the cTPQ state. The detailed calculation will be shown in Sect. 3.7. In Chap. 2, we do not have any formula to obtain the genuine thermodynamic variables from a single realization of $|\psi_E\rangle$. As the GTVs are related to the number of the microstates, one may think that it is impossible to evaluate from a single pure state. However, since $|\beta, V\rangle$ is an unnormalized vector, I will show that the normalization constant contains the information of the density of state, and thus, the free energy is obtained from the normalization of the cTPQ state,

$$f(\beta; V) = -\frac{1}{V\beta} \ln\langle\beta, V|\beta, V\rangle, \tag{3.10}$$

with the stochastic error getting exponentially small as V increasing. Here, $f(\beta; V)$ is the free energy density, $f(\beta; V) \equiv \frac{F(\beta,V)}{V}$. This stochastic error is the error coming from the random variables $\{z_i\}_i$. I estimate this error and get

$$\overline{\left(\langle\beta, V|\beta, V\rangle - \overline{\langle\beta, V|\beta, V\rangle}\right)^2} = Z(2\beta, V). \tag{3.11}$$

In order see that the variance is negligibly small as compared to the average, I divide the variance by the average:

$$\begin{aligned} \overline{\left(\frac{\langle\beta, V|\beta, V\rangle}{\overline{\langle\beta, V|\beta, V\rangle}} - 1\right)^2} &= \frac{Z(2\beta, V)}{Z(\beta, V)^2} \\ &= \frac{1}{\exp\left[2V\beta\left(f(\frac{1}{2\beta}; V) - f(\frac{1}{\beta}; V)\right)\right]}. \end{aligned} \tag{3.12}$$

Since $f(\frac{1}{2\beta}; V) - f(\frac{1}{\beta}; V) \geq 0$ and $\Theta(V^0)$, the rhs of Eq. (3.12) is $\exp[-\Theta(V)]$. This result is rewritten by using a Markov-type inequality. The inequality is as follows: Let x be a real random variable and y a real number, then for arbitrary $\epsilon > 0$,

$$\mathrm{P}\left(|x - y| \geq \epsilon\right) \leq \overline{(x - y)^2}/\epsilon^2, \tag{3.13}$$

Using this inequality, I show that the length of the cTPQ state rarely deviate from its average, $Z(\beta, V)$. For any positive constant ϵ,

$$\mathrm{P}\left(\left|\frac{\langle\beta, V|\beta, V\rangle}{\overline{\langle\beta, V|\beta, V\rangle}} - 1\right| \geq \epsilon\right) \leq \frac{1}{\epsilon^2 \exp[2V\beta\{f(1/2\beta; V) - f(1/\beta; V)\}]}. \tag{3.14}$$

This inequality indicates that the probability that $\langle\beta, V|\beta, V\rangle / \overline{\langle\beta, V|\beta, V\rangle}$ deviates from 1 is exponentially small. Thus, from a single $\langle\beta, V|\beta, V\rangle$ gives the free energy correctly within the exponentially small error.

Once we get the thermodynamic functions, any equilibrium values of the MVs and the GTVs are obtained from the derivatives of them. However, the ensemble averages are practically more useful to get the equilibrium values of MVs in the ensemble formulation. In the TPQ formulation, the equilibrium values are obtained from the expectation values of MVs using the TPQ states, i.e.,

$$\langle \hat{A} \rangle_{\beta,V}^{\mathrm{TPQ}} = \langle \hat{A} \rangle_{\beta,V}^{\mathrm{ens}} \tag{3.15}$$

with the error being exponentially small. Here,

$$\langle \hat{A} \rangle_{\beta,V}^{\mathrm{TPQ}} \equiv \frac{\langle \beta, V | \hat{A} | \beta, V \rangle}{\langle \beta, V | \beta, V \rangle}. \tag{3.16}$$

Just like the ensemble average, calculating the expectation value is more useful than calculating the derivative of the thermodynamic function. I estimate the upper bound of the error of Eq. (3.15). Dropping smaller order terms, I get

$$\overline{(\langle \hat{A} \rangle_{\beta,V}^{\mathrm{TPQ}} - \langle \hat{A} \rangle_{\beta,V}^{\mathrm{ens}})^2} \leq \frac{\langle (\Delta \hat{A})^2 \rangle_{2\beta,V}^{\mathrm{ens}} + (\langle A \rangle_{2\beta,V}^{\mathrm{ens}} - \langle A \rangle_{\beta,V}^{\mathrm{ens}})^2}{\exp[2V\beta\{f(1/2\beta;V) - f(1/\beta;V)\}]}, \tag{3.17}$$

where $\langle (\Delta \hat{A})^2 \rangle_{\beta,V}^{\mathrm{ens}} \equiv \langle (\hat{A} - \langle A \rangle_{\beta,V}^{\mathrm{ens}})^2 \rangle_{\beta,V}^{\mathrm{ens}}$. Using inequality (3.13), I get

$$\mathrm{P}\left(\left| \langle \hat{A} \rangle_{\beta,V}^{\mathrm{TPQ}} - \langle \hat{A} \rangle_{\beta,V}^{\mathrm{ens}} \right| \geq \epsilon \right) \leq \frac{1}{\epsilon^2} \frac{\langle (\Delta \hat{A})^2 \rangle_{2\beta,V}^{\mathrm{ens}} + (\langle A \rangle_{2\beta,V}^{\mathrm{ens}} - \langle A \rangle_{\beta,V}^{\mathrm{ens}})^2}{\exp[2V\beta\{f(T/2,\mu;V) - f(T,;V)\}]}. \tag{3.18}$$

Since the rhs of inequality (3.18) is smaller than $\frac{V^{2m}}{\exp[\Theta(V)]}$, inequality (3.18) means that the expectation value of cTPQ state is very close to the ensemble average. This result is similar to inequality (2.32) of $|\psi_E\rangle$. However, the rhs of inequality (3.18) is controlled not by entropy but by the free energy. This is a consequence of not using the TPQ state whose energy is specified but using the TPQ state whose temperature is specified.

To sum up,

Genuine Thermodynamic Variables
The norm of a single realization of the cTPQ state gives the free energy. Its error is exponentially small as shown in inequality (3.14). From the free energy, any equilibrium values of genuine thermodynamic variables are obtained.
Mechanical Variables
For any $\hat{A} \in$ MVs, the expectation value of a single realization of the cTPQ state gives the equilibrium value correctly. Its error is exponentially small as shown in inequality (3.18)

I have shown that a single realization of the cTPQ state is sufficient to evaluate *any* quantities which are interesting in statistical mechanics. Therefore, a new formulation of statistical mechanics based on the TPQ state is established.

I notice that All the terms in the rhs of inequalities (3.14) and (3.18) can be estimated by the cTPQ state itself. Thus, the cTPQ state evaluates its error from its own results. This self-validating property is useful in practical applications.

3.4 Mechanism

If the cTPQ state is regarded as an equilibrium state, its energy density converges to some value. I will show that the cTPQ state has a sharp-peaked energy distribution in the energy shell. In this subsection, I assume the dimension of the Hilbert space $\dim\mathscr{H}$ is finite for the simplicity. Even when $\dim\mathscr{H}$ is infinite, the calculation in this subsection will be easily extended. When $\dim\mathscr{H} < \infty$, the cTPQ state is rewritten as

$$|\beta, V\rangle = \exp\left[-\frac{\beta\hat{H}}{2}\right]|\psi_0\rangle, \tag{3.19}$$

because the norm of $|\psi_0\rangle$ is finite and thus $|\psi_0\rangle$ is well defined. Here,

$$|\psi_0\rangle = \sum_n z_i|i\rangle. \tag{3.20}$$

Then, I expand $|\psi_0\rangle$ by energy eigenstates basis:

$$|\psi_0\rangle = \sum_n z_n|n\rangle \tag{3.21}$$

where $|n\rangle$ is an energy eigenvector such that $\hat{h}|n\rangle = u_n|n\rangle$, I notice that random variables z_n have an invariant probability measure under any unitary transformations. Therefore, $\{z_n\}_n$ has the same probability measure as $\{z_i\}_i$.

I define an energy distribution $\tilde{g}(u)$ as

$$\tilde{g}(u) \equiv \sum_{n\in\mathscr{S}(u)} |z_n|^2 \tag{3.22}$$

where $\mathscr{S}(u) = \{n|u_n \in (u-\delta, u]\}$. Although each $|z_n|^2$ takes random value, the law of large numbers works for the sum $\tilde{g}(u)$, because $\dim\mathscr{S}(u) = \exp[Ns(u; V)] = \exp[\Theta(N)]$ at finite temperature. Therefore, the sum $\tilde{g}(u)$ converges to the density of states, i.e.,

$$\tilde{g}(u) \simeq g(u). \tag{3.23}$$

For the cTPQ state $|\beta, V\rangle = \exp[-\frac{\beta\hat{H}}{2}]|\psi_0\rangle$, the energy distribution $h(u)$ is

$$h(u) \equiv \sum_{n\in\mathscr{S}(u)} |z_n|^2 \exp[-V\beta u] \tag{3.24}$$

$$\simeq g(u)\exp[-V\beta u] \tag{3.25}$$

$$= \exp\left[V\left(s(u;V) - \beta u\right)\right]. \tag{3.26}$$

The rhs of Eq. (3.26) is the same as the energy distribution of the Gibbs state at the same temperature. Since the rhs of Eq. (3.26) has a peak at u^* such that

$$\left.\frac{\partial s(u;V)}{\partial u}\right|_{u=u^*} = \beta, \tag{3.27}$$

I expand the rhs around $u = u^*$ and get.

$$h(u) \simeq \exp\left[V\left(s(u^*;V) + \frac{\partial^2 s(u^*;V)}{\partial u^2}(u-u^*)^2 + O\left((u-u^*)^3\right)\right)\right]. \tag{3.28}$$

When the system is consistent with thermodynamics, entropy has to be a convex function of energy, i.e.,

$$\frac{\partial^2 s(u^*;V)}{\partial u^2} \leq 0, \tag{3.29}$$

at least for sufficient large V, and entropy is extensive, i.e.,

$$s(u^*;V) = s(u^*) + o(V^0), \tag{3.30}$$

where $s(u^*)$ is the entropy density for $V \to \infty$ and is the order of unity. Therefore, Eq. (3.28) means that $h(u)$ has a sharp peak at $u = u^*$ and the width of the peak is $\Theta(\frac{1}{\sqrt{V}})$. Since the energy distribution of the cTPQ state is concentrated on the target energy u^*, I can expect that the cTPQ state for large V will be regarded as the equilibrium state at this energy. Namely, the cTPQ state and $|\psi_E\rangle$ become almost the same state at large V.

3.5 Thermo Field Dynamics

I also note that the TPQ states are completely different from "purification" [5] of a density matrix such as thermo field dynamics (TFD) [6]. The purification is the technique that every density matrix can be represented as a pure quantum state by introducing ancilla's degrees of freedom. In order to prove that the purification is always available, I make an example of the purification for an arbitrary density matrix,

$$\hat{\rho} = \sum_i w_i |i\rangle_1 \langle i|_1 \tag{3.31}$$

where $\{w_i\}_i$ is a set of probability with $\sum_i w_i = 1$ and $\{|i\rangle_1\}_i$ is an orthonormal basis in $\mathscr{H}_1$. Let $\mathscr{H}_2$ be a copy of $\mathscr{H}_1$. Then, I define

$$|\psi\rangle \in \mathscr{H}_1 \otimes \mathscr{H}_2 \tag{3.32}$$

by

$$|\psi\rangle \equiv \sum_i \sqrt{w_i} |i\rangle_1 \otimes |i\rangle_2 \tag{3.33}$$

Obviously, by tracing out $\mathscr{H}_2$ from $|\psi\rangle$, I get

$$\mathrm{Tr}_B |\psi\rangle\langle\psi| = \hat{\rho} \tag{3.34}$$

This "purified" pure state gives identical results to the density matrix. In the TFD, one does not employ an orthonormal basis as $\{|i\rangle_1\}_i$, but employs $\{|i; \beta, V\rangle_1\}_i$ defined in Eq. (3.2). Then, the TFD reads

$$|\psi_{\mathrm{TFD}}\rangle \equiv \sum_i \frac{|i; \beta, V\rangle_1}{Z(\beta, V)} \otimes |i\rangle_2. \tag{3.35}$$

The TFD use the ancilla and exactly equivalent to the Gibbs state.

In contrast to this, the TPQ states do not use any ancillae. Since the corresponding density matrix such as the Gibbs state is mixed, a single realization of the TPQ states are quantum mechanically completely different from it. Moreover, different realizations of the TPQ states are microscopically very different from each other. However, thanks to a vast number of a degrees of freedom of a Hilbert space, almost all the TPQ states are *macroscopically* identical to each other. They are also equivalent to the corresponding density matrix macroscopically. This findings are the heart of the TPQ state.

3.6 Grandcanonical Thermal Pure Quantum State

In light of the discussion of the cTPQ state, the TPQ state corresponding to the grandcanonical ensemble is easily defined [2, 4]. I call it a grandcanonical TPQ (gTPQ) state. Firstly, I begin with

$$|i; \beta, \mu, V\rangle \equiv \exp[-\frac{1}{2}\beta(\hat{H} - \mu\hat{N})]|i\rangle, \tag{3.36}$$

where $\{|i\rangle\}_i$ is an arbitrary orthonormal basis, and $\hat{N}$ is the number operator. Then, I superpose $|\nu; \beta, \mu, V\rangle$'s as

$$|\beta, \mu, V\rangle \equiv \sum_i z_i |i; \beta, \mu, V\rangle. \tag{3.37}$$

$|\beta, \mu, V\rangle$ is the gTPQ state. All the results in Sect. 3.3 hold for the gTPQ state by substituting

$$Z(\beta, V) \to \Xi(\beta, \nu, V) \equiv \mathrm{Tr}[\exp\{-\beta(\hat{H} - \nu\hat{N})\}] \tag{3.38}$$

$$f(\beta, V) \to j(\beta, \nu, V) \equiv \frac{1}{\beta V} \ln \Xi(\beta, \nu, V) \tag{3.39}$$

$$\langle A\rangle^{\mathrm{ens}}_{\beta,V} \to \langle A\rangle^{\mathrm{ens}}_{\beta,\mu,V} \tag{3.40}$$

The gTPQ state is introduced and discussed in Ref. [4].

3.7 Derivations of Main Results

In this subsection, I introduce the main results, formula (3.12) and (3.18).

3.7.1 *Random Average and Variance*

I consider two random variables $f = f_0 + \delta f$ and $g = g_0 + \delta g$, where f_0 and g_0 are their mean and δf and δg are their corresponding random variable, respectively. I calculate an average and variance of f/g. First, I expand f/g as follows:

$$\frac{f_0 + \delta f}{g_0 + \delta g} = (f_0 + \delta f)\frac{1}{g_0}\left(1 - \frac{\delta g}{g_0} + \frac{\delta g^2}{g_0^2} - \cdots\right) \tag{3.41}$$

$$= \frac{f_0}{g_0} + \frac{\delta f}{g_0} - \frac{f_0 \delta g}{g_0^2} - \frac{\delta f \delta g}{g_0^2} + \frac{f_0 \delta g^2}{g_0^3} + O(\delta^3) \tag{3.42}$$

So, the average of f/g is

$$\overline{\left(\frac{f_0 + \delta f}{g_0 + \delta g}\right)} = \frac{f_0}{g_0} - \frac{\overline{\delta f \delta g}}{g_0^2} + \frac{f_0 \overline{\delta g^2}}{g_0^3} + O(\overline{\delta^3}) \tag{3.43}$$

and the variance is

$$\overline{\left(\frac{f_0+\delta f}{g_0+\delta g}-\overline{\left(\frac{f_0+\delta f}{g_0+\delta g}\right)}\right)^2}=\overline{\left(\frac{\delta f}{g_0}-\frac{f_0\delta g}{g_0^2}\right)^2}+O(\overline{\delta^3}) \tag{3.44}$$

$$=\frac{\overline{\delta f^2}}{g_0^2}-2\frac{f_0\overline{\delta f\delta g}}{g_0^3}+\frac{f_0^2\overline{\delta g^2}}{g_0^4}+O(\overline{\delta^3}) \tag{3.45}$$

In order to calculate an average and variance of the form f/g, I need three terms, $\overline{\delta f^2}$, $\overline{\delta f\delta g}$, and $\overline{\delta g^2}$. In the following subsections, I calculate these terms.

Before starting the calculation of respective problems in the following subsections, I derive a formula of a general form. I calculate

$$\overline{\left(\langle\psi_0|\hat{A}|\psi_0\rangle-\overline{\langle\psi_0|\hat{A}|\psi_0\rangle}\right)\left(\langle\psi_0|\hat{B}|\psi_0\rangle-\overline{\langle\psi_0|\hat{B}|\psi_0\rangle}\right)} \tag{3.46}$$

At first, $\{z_n\}_n$ satisfies

$$\overline{z_n}=0 \tag{3.47}$$

$$\overline{z_n^*z_m}=\delta_{n,m} \tag{3.48}$$

$$\overline{|z_n|^4}=2 \tag{3.49}$$

$$\overline{|z_n|^2|z_m|^2}=1\quad(n\neq m) \tag{3.50}$$

$$\overline{z_k^*z_l^*z_mz_n}=0\quad\text{(except for above two terms)} \tag{3.51}$$

Then, I get

$$\overline{\left(\langle\psi_0|\hat{A}|\psi_0\rangle-\overline{\langle\psi_0|\hat{A}|\psi_0\rangle}\right)\left(\langle\psi_0|\hat{B}|\psi_0\rangle-\overline{\langle\psi_0|\hat{B}|\psi_0\rangle}\right)} \tag{3.52}$$

$$=\sum_{n,m,n',m'}\overline{z_n^*z_mz_{n'}^*z_{m'}}\langle n|\hat{A}|m\rangle\langle n'|\hat{B}|m'\rangle$$
$$-\sum_{n,m}\overline{z_n^*z_m}\langle n|\hat{A}|m\rangle\sum_{n',m'}\overline{z_{n'}^*z_{m'}}\langle n'|\hat{B}|m'\rangle \tag{3.53}$$

$$=\sum_n\overline{|z_n|^4}\langle n|\hat{A}|n\rangle\langle n|\hat{B}|n\rangle+\sum_{n\neq n'}\overline{|z_n|^2|z_{n'}|^2}\langle n|\hat{A}|n\rangle\langle n'|\hat{B}|n'\rangle$$
$$+\sum_{n\neq m}\overline{|z_n|^2|z_m|^2}\langle n|\hat{A}|m\rangle\langle m|\hat{B}|n\rangle-\sum_n\overline{|z_n|^2}\langle n|\hat{A}|n\rangle\sum_{n'}\overline{|z_{n'}|^2}\langle n'|\hat{B}|n'\rangle$$

$$=2\sum_n\langle n|\hat{A}|n\rangle\langle n|\hat{B}|n\rangle+\sum_{n\neq m}\langle n|\hat{A}|n\rangle\langle m|\hat{B}|m\rangle$$
$$+\sum_{n\neq m}\langle n|\hat{A}|m\rangle\langle m|\hat{B}|n\rangle-\sum_{n,m}\langle n|\hat{A}|n\rangle\langle m|\hat{B}|m\rangle \tag{3.54}$$

$$=\sum_{n,m}\langle n|\hat{A}|m\rangle\langle m|\hat{B}|n\rangle \tag{3.55}$$

3.7.2 Normalization Constant (Partition Function)

Now, I prove that the canonical TPQ state $|\beta, V\rangle$ gives the correct equilibrium values by calculating the corresponding expectation values. Firstly, I see $\langle\beta, V|\beta, V\rangle$, which works like a partition function. Its average behaves as

$$\overline{\langle\beta, V|\beta, V\rangle} = \frac{1}{D}\sum_n \exp(-V\beta u_n) \tag{3.56}$$

$$= Z(\beta, V). \tag{3.57}$$

Then, I see its variance.

$$\overline{(\langle\beta, V|\beta, V\rangle - \overline{\langle\beta, V|\beta, V\rangle})^2} = \overline{\langle\beta, V|\beta, V\rangle^2} - (\overline{\langle\beta, V|\beta, V\rangle})^2 \tag{3.58}$$

In Eq. (3.55), this is the case of $\hat{A} = \hat{B} = \exp[-V\beta\hat{h}]$. Thus, I get

$$\overline{(\langle\beta, V|\beta, V\rangle - \overline{\langle\beta, V|\beta, V\rangle})^2} = \sum_n \langle n|\exp[-2V\beta\hat{h}]|n\rangle \tag{3.59}$$

$$= \sum_n \exp[-2V\beta u_n] \tag{3.60}$$

$$= Z(2\beta, V) \tag{3.61}$$

Thus, I get Eq. (3.12).

3.7.3 Mechanical Variables

Then, I consider an expectation value of mechanical variable, i.e.,

$$\frac{\langle\beta, V|\hat{M}|\beta, V\rangle}{\langle\beta, V|\beta, V\rangle}. \tag{3.62}$$

Using Eqs. (3.43) and (3.45), I find that I need to calculate these two terms,

$$\overline{\left(\langle\beta, V|\hat{M}|\beta, V\rangle - \overline{\langle\beta, V|\hat{M}|\beta, V\rangle}\right)^2} \tag{3.63}$$

$$\overline{\left(\langle\beta, V|\hat{M}|\beta, V\rangle - \overline{\langle\beta, V|\hat{M}|\beta, V\rangle}\right)\left(\langle\beta, V|\beta, V\rangle - \overline{\langle\beta, V|\beta, V\rangle}\right)} \tag{3.64}$$

First, I calculate the term (3.63). Using Eq. (3.55) with $\hat{A} = \hat{B} = \exp[-\frac{1}{2}V\beta\hat{h}]\hat{M}$ $\exp[-\frac{1}{2}V\beta\hat{h}]$, I get

$$\overline{\left(\langle\beta, V|\hat{M}|\beta, V\rangle - \overline{\langle\beta, V|\hat{M}|\beta, V\rangle}\right)^2}$$
$$= \sum_n \langle n|\exp[-\frac{1}{2}V\beta\hat{h}]\hat{M}\exp[-V\beta\hat{h}]\hat{M}\exp[-\frac{1}{2}V\beta\hat{h}]|n\rangle \tag{3.65}$$
$$= \sum_n \exp(-V\beta u_n)\langle n|\hat{M}\exp(-V\beta\hat{h})\hat{M}|n\rangle. \tag{3.66}$$

Then, I calculate the term (3.64). Using Eq. (3.55) with $\hat{A} = \exp[-\frac{1}{2}V\beta\hat{h}]\hat{M}\exp[-\frac{1}{2}V\beta\hat{h}]$ and $\hat{B} = \exp[-V\beta\hat{h}]$, I get

$$\overline{\left(\langle\beta, V|\hat{M}|\beta, V\rangle - \overline{\langle\beta, V|\hat{M}|\beta, V\rangle}\right)\left(\langle\beta, V|\beta, V\rangle - \overline{\langle\beta, V|\beta, V\rangle}\right)}$$
$$= \sum_n \langle n|\exp[-\frac{1}{2}V\beta\hat{h}]\hat{M}\exp[-\frac{3}{2}V\beta\hat{h}]|n\rangle \tag{3.67}$$
$$= \sum_n \langle n|\hat{M}|n\rangle\exp(-2V\beta u_n) \tag{3.68}$$
$$= \langle\hat{M}\rangle^{\mathrm{ens}}_{2\beta,V}Z(2\beta, V) \tag{3.69}$$

Now, I am ready to evaluate the average and the variance of $\langle\beta, V|\hat{M}|\beta, V\rangle/\langle\beta, V|\beta, V\rangle$.

(i) mean

Using Eqs. (3.43), (3.61), and (3.69) and neglecting $O(\delta^4)$ terms in Eq. (3.43), I get the average

$$\left|\overline{\left(\frac{\langle\beta, V|\hat{M}|\beta, V\rangle}{\langle\beta, V|\beta, V\rangle}\right)} - \frac{\overline{\langle\beta, V|\hat{M}|\beta, V\rangle}}{\overline{\langle\beta, V|\beta, V\rangle}}\right|$$
$$= \left|-\frac{\langle\hat{M}\rangle^{\mathrm{ens}}_{2\beta,V}Z(2\beta, V)}{\overline{\langle\beta, V|\beta, V\rangle}^2} + \frac{Z(2\beta, V)\overline{\langle\beta, V|\hat{M}|\beta, V\rangle}}{\overline{\langle\beta, V|\beta, V\rangle}^3}\right| \tag{3.70}$$
$$= \left|-\frac{\langle\hat{M}\rangle^{\mathrm{ens}}_{2\beta,V}Z(2\beta, V)}{(Z(\beta))^2} + \frac{\langle\hat{M}\rangle^{\mathrm{ens}}_{\beta,V}Z(2\beta, V)}{(Z(\beta))^2}\right| \tag{3.71}$$
$$= \left|-\frac{\left(\langle\hat{M}\rangle^{\mathrm{ens}}_{2\beta,V} - \langle\hat{M}\rangle^{\mathrm{ens}}_{\beta,V}\right)Z(2\beta, V)}{(Z(\beta))^2}\right| \tag{3.72}$$
$$= \left|\frac{Z(2\beta, V)\left(\langle\hat{M}\rangle^{\mathrm{ens}}_{2\beta,V} - \langle\hat{M}\rangle^{\mathrm{ens}}_{\beta,V}\right)}{Z(\beta, V)^2}\right| \tag{3.73}$$
$$= \frac{\left|\langle\hat{M}\rangle^{\mathrm{ens}}_{2\beta,V} - \langle\hat{M}\rangle^{\mathrm{ens}}_{\beta,V}\right|}{\exp[2V\beta(f(1/2\beta; V) - f(1/\beta; V))]} \tag{3.74}$$

(ii) variance : a rough estimate
Using Eqs. (3.45), (3.61), (3.66), and (3.69), I get the variance

$$\overline{\left(\frac{\langle\beta,V|\hat{M}|\beta,V\rangle}{\langle\beta,V|\beta,V\rangle}-\overline{\left(\frac{\langle\beta,V|\hat{M}|\beta,V\rangle}{\langle\beta,V|\beta,V\rangle}\right)}\right)^2}$$

$$=\frac{\sum_n \exp(-V\beta u_n)\langle n|\hat{M}\exp(-V\beta\hat{h})\hat{M}|n\rangle}{\overline{\langle\beta,V|\beta,V\rangle}^2}$$

$$-2\frac{\langle\hat{M}\rangle^{\text{ens}}_{\beta,V}\langle\hat{M}\rangle^{\text{ens}}_{2\beta,V}Z(2\beta,V)}{\overline{\langle\beta,V|\beta,V\rangle}^2}+\frac{{\langle\hat{M}\rangle^{\text{ens}}_{\beta,V}}^2 Z(2\beta,V)}{\overline{\langle\beta,V|\beta,V\rangle}^2} \tag{3.75}$$

$$=\frac{1}{Z(\beta)^2}\left(\left\{\sum_n \exp(-V\beta u_n)\langle n|\hat{M}\exp(-V\beta\hat{h})\hat{M}|n\rangle\right\}\right.$$

$$\left.-2\langle\hat{M}\rangle^{\text{ens}}_{\beta,V}\langle\hat{M}\rangle^{\text{ens}}_{2\beta,V}Z(2\beta,V)+{\langle\hat{M}\rangle^{\text{ens}}_{\beta,V}}^2 Z(2\beta,V)\right) \tag{3.76}$$

This terms are estimated roughly as follows:

$$\text{Eq. (3.76)} =\frac{1}{Z(\beta)^2}\sum_n$$

$$\left(\left(\sum_{m\neq n}|\langle n|\hat{M}|m\rangle|^2\exp(-V\beta(u_n+u_m)\right)\right.$$

$$\left.+\left(\langle n|\hat{M}|n\rangle-\langle\hat{M}\rangle^{\text{ens}}_{\beta,V}\right)^2\exp(-2V\beta u_n)\right)$$

$$\leq\frac{1}{Z(\beta)^2}\sum_n \exp(-V\beta u_n)\exp(-V\beta u_{\min})$$

$$\left(\left(\sum_{m\neq n}|\langle n|\hat{M}|m\rangle|^2\right)+\left(\langle n|\hat{M}|n\rangle-\langle\hat{M}\rangle^{\text{ens}}_{\beta,V}\right)^2\right) \tag{3.77}$$

$$=\frac{\exp(-V\beta u_{\min})}{Z(\beta,V)}\left(\sum_n\langle\hat{M}^2\rangle^{\text{ens}}_{\beta,V}-{\langle\hat{M}\rangle^{\text{ens}}_{\beta,V}}^2\right)$$

$$=\frac{\langle(\Delta\hat{M})^2\rangle^{\text{ens}}_{\beta,V}}{\exp[V\beta(f(0;V)-f(1/\beta;V))]}. \tag{3.78}$$

From this result, I get

$$\text{Prob}\left(\left|\frac{\langle\beta, V|\hat{M}|\beta, V\rangle}{\langle\beta, V|\beta, V\rangle} - \overline{\left(\frac{\langle\langle\beta, V|\hat{M}|\beta, V\rangle}{\langle\beta, V|\beta, V\rangle}\right)}\right| \geq \epsilon\right)$$
$$\leq \frac{1}{\epsilon^2}\frac{\langle(\Delta\hat{M})^2\rangle^{\text{ens}}_{\beta,V}}{\exp[V\beta(f(0;V) - f(1/\beta;V))]}. \tag{3.79}$$

Note that the rhs vanishes when

$$\langle(\Delta\hat{M})^2\rangle^{\text{ens}}_{\beta,V} = 0. \tag{3.80}$$

That is, the canonical TPQ state always gives the exact results for mechanical variables that satisfies this condition.

(iii) variance : a better estimate

To get a better estimate, I go back to Eq. (3.76). If I number u_n such that $u_n \leq e_{n'}$ for $n < n'$, I have

$$\{\} \text{ in Eq. (3.76)} = \sum_{m,n} |\langle n|\hat{M}|m\rangle|^2 e^{-V\beta(u_n+u_m)} \tag{3.81}$$
$$\leq \sum_{m,n} |\langle n|\hat{M}|m\rangle|^2 \frac{e^{-2V\beta u_m} + e^{-2V\beta u_n}}{2} \tag{3.82}$$
$$= \frac{1}{2}\sum_m \langle m|\hat{M}^2|m\rangle e^{-2V\beta u_m} + \frac{1}{2}\sum_n \langle n|\hat{M}^2|n\rangle e^{-2V\beta u_n}$$
$$= \langle\hat{M}^2\rangle^{\text{ens}}_{2\beta,V} Z(2\beta). \tag{3.83}$$

Therefore,

$$\text{Eq. (3.76)} \leq \frac{Z(2\beta)}{Z(\beta)^2}\left(\langle\hat{M}^2\rangle^{\text{ens}}_{2\beta,V} - 2\langle\hat{M}\rangle^{\text{ens}}_{\beta,V}\langle\hat{M}\rangle^{\text{ens}}_{2\beta,V} + \langle\hat{M}\rangle^{\text{ens}\,2}_{\beta,V}\right). \tag{3.84}$$

I can rewrite this by noting that the lhs does not depend on the origin of $\hat{M}$ [because it vanishes when $\hat{M} = \hat{1}$] as

$$\text{Eq. (3.76)} = \text{Eq. (3.76) with } \hat{M} \to \hat{M} - \langle\hat{M}\rangle^{\text{ens}}_{2\beta,V} \tag{3.85}$$
$$\leq \frac{Z(2\beta)}{Z(\beta)^2}\left(\langle(\hat{M} - \langle\hat{M}\rangle^{\text{ens}}_{2\beta,V})^2\rangle^{\text{ens}}_{2\beta,V} + [\langle\hat{M}\rangle^{\text{ens}}_{\beta,V} - \langle\hat{M}\rangle^{\text{ens}}_{2\beta,V}]^2\right) \tag{3.86}$$
$$= \frac{\langle(\Delta\hat{M})^2\rangle^{\text{ens}}_{2\beta,V} + [\langle\hat{M}\rangle^{\text{ens}}_{\beta,V} - \langle\hat{M}\rangle^{\text{ens}}_{2\beta,V}]^2}{\exp[2V\beta(f(1/2\beta;V) - f(1/\beta;V))]}, \tag{3.87}$$

where

$$\langle (\Delta \hat{M})^2 \rangle^{\text{ens}}_{2\beta,V} \equiv \langle (\hat{M} - \langle \hat{M} \rangle^{\text{ens}}_{2\beta,V})^2 \rangle^{\text{ens}}_{2\beta,V}. \tag{3.88}$$

From this result, I get

$$\text{Prob}\left(\left| \frac{\langle \beta, V | \hat{M} | \beta, V \rangle}{\langle \beta, V | \beta, V \rangle} - \overline{\left(\frac{\langle \langle \beta, V | \hat{M} | \beta, V \rangle}{\langle \beta, V | \beta, V \rangle} \right)} \right| \geq \epsilon \right)$$
$$\leq \frac{1}{\epsilon^2} \frac{\langle (\Delta \hat{M})^2 \rangle^{\text{ens}}_{2\beta,V} + [\langle \hat{M} \rangle^{\text{ens}}_{\beta,V} - \langle \hat{M} \rangle^{\text{ens}}_{2\beta,V}]^2}{\exp[2V\beta(f(1/2\beta; V) - f(1/\beta; V))]}. \tag{3.89}$$

Note that the rhs vanishes when

$$\langle (\Delta \hat{M})^2 \rangle^{\text{ens}}_{2\beta,V} = 0 \text{ and } \langle \hat{M} \rangle^{\text{ens}}_{\beta,V} = \langle \hat{M} \rangle^{\text{ens}}_{2\beta,V}. \tag{3.90}$$

That is, the canonical TPQ state always gives the exact results for mechanical variables that satisfies this condition.

When conditions neither (3.80) nor (3.90) is satisfied, I have

$$\begin{aligned} \frac{\text{rhs of (3.89)}}{\text{rhs of (3.79)}} &= \frac{\langle (\Delta \hat{M})^2 \rangle^{\text{ens}}_{2\beta,V} + [\langle \hat{M} \rangle^{\text{ens}}_{\beta,V} - \langle \hat{M} \rangle^{\text{ens}}_{2\beta,V}]^2}{\langle (\Delta \hat{M})^2 \rangle^{\text{ens}}_{\beta,V}} \\ &\quad \times \exp[V\beta\{f(0; V) - f(1/\beta; V)\} - 2V\beta\{f(1/2\beta; V) - f(1/\beta; V)\}] \\ &= \frac{\langle (\Delta \hat{M})^2 \rangle^{\text{ens}}_{2\beta,V} + [\langle \hat{M} \rangle^{\text{ens}}_{\beta,V} - \langle \hat{M} \rangle^{\text{ens}}_{2\beta,V}]^2}{\langle (\Delta \hat{M})^2 \rangle^{\text{ens}}_{\beta,V}} \\ &\quad \times \exp\left[-V\beta\left(\{f(\frac{1}{2\beta}; V) - f(\frac{1}{\beta}; V)\} - \{f(0; V) - f(\frac{1}{2\beta}; V)\}\right)\right]. \end{aligned}$$

Since $f(T; V)$ is convex up as a function of T $(= 1/\beta)$ and $\partial f / \partial T = -s < 0$ for all T,

$$\{f(1/2\beta; V) - f(1/\beta; V)\} - \{f(0; V) - f(1/2\beta; V)\} = \Theta(1). \tag{3.91}$$

Therefore,

$$\frac{\text{RHS of (3.89)}}{\text{RHS of (3.79)}} = \frac{\langle (\Delta \hat{M})^2 \rangle^{\text{ens}}_{2\beta,V} + [\langle \hat{M} \rangle^{\text{ens}}_{\beta,V} - \langle \hat{M} \rangle^{\text{ens}}_{2\beta,V}]^2}{\langle (\Delta \hat{M})^2 \rangle^{\text{ens}}_{\beta,V}} \exp[-V\beta\Theta(1)]. \tag{3.92}$$

When the reasonable condition

$$\frac{\langle(\Delta\hat{M})^2\rangle^{\text{ens}}_{2\beta,V} + [\langle\hat{M}\rangle^{\text{ens}}_{\beta,V} - \langle\hat{M}\rangle^{\text{ens}}_{2\beta,V}]^2}{\langle(\Delta\hat{M})^2\rangle^{\text{ens}}_{\beta,V}} \le \text{constant} \times N^m, \quad m = o(N), \tag{3.93}$$

is satisfied, I have

$$\frac{\text{RHS of (3.89)}}{\text{RHS of (3.79)}} \le \text{constant} \times N^m \exp[-V\beta\Theta(1)] \le \exp[-V\beta\Theta(1) + o(N)\ln N]. \tag{3.94}$$

This indicates that rhs of (3.79) gets exponentially smaller than rhs of (3.79) with increasing N. Since condition (3.93) seems to be satisfied for most mechanical variables of interest, I conclude that inequality (3.89) is better than inequality (3.79) for most cases of interest.

References

1. S. Sugiura, A. Shimizu, Phys. Rev. Lett. **108**, 240401 (2012)
2. S. Sugiura, A. Shimizu, Phys. Rev. Lett. **111**, 010401 (2013)
3. H.B. Callen, *Thermodynamics* (John Wiley and Sons, New York, 1960)
4. M. Hyuga, S. Sugiura, K. Sakai, and A. Shimizu Phys. Rev. B **90**, 121110(R) (2014)
5. M.A. Nielsen, I.L. Chuang, *Quantum Computation and Quantum Information* (Cambridge University Press, Cambridge, 2000)
6. Y. Takahashi, H. Umezawa, Int. J. Mod. Phys. B **10**, 1755 (1996)

Chapter 4
Microcanonical Thermal Pure Quantum State

In this section, I introduce a microcanonical TPQ (mTPQ) state [1]. The mTPQ state is the state which aims at practical applications. Although there are many possibilities to construct the TPQ state with energy being specified, I believe the mTPQ state is one of the most efficient TPQ states to generate in the applications.

4.1 Construction of TPQ State

Since I consider the state which is easy to generate in the practical applications, I restrict $\dim\mathscr{H} < \infty$. Although the cTPQ state is well defined even in the case of $\dim\mathscr{H} = \infty$, I usually truncate high energy states and make $\dim\mathscr{H} < \infty$ in practical applications.

Thanks to the restriction to $\dim\mathscr{H} < \infty$, I can start from a random vector in the Hilbert state,

$$|\psi_0\rangle = \sum_i z_i |i\rangle, \tag{4.1}$$

where $z_i \equiv \frac{(x_i + y_i)}{\sqrt{2}}$ is a random variable with probability $\mathrm{P}(x_i = a) = \mathrm{P}(y_i = a) = \frac{1}{\sqrt{2\pi}} \exp[-\frac{a^2}{2}]$, and $\{i\}_i$ is an arbitrary orthonormal basis set spanning the Hilbert space.

Then, I multiply some polynomials of Hamiltonian to it and make its energy distribution be concentrated to a target energy density u_0. There are many candidates of the multiplying polynomials of Hamiltonian. Here, I introduce three of them.

(1)

$$|\psi_1(u_0, k)\rangle \equiv \{1 - \alpha(\hat{h} - u_0)^2\}^k |\psi_0\rangle \tag{4.2}$$

S. Sugiura, *Formulation of Statistical Mechanics Based on Thermal Pure Quantum States*, Springer Theses, DOI 10.1007/978-981-10-1506-9_4

where $\hat{h} \equiv \hat{H}/V$ and α is an arbitrary parameter such that $\frac{1}{\|\hat{h}-u_0\|^2} \geq \alpha$. For sufficiently large k, $|\psi_1(u_0, k)\rangle$ has a sharp peak around u_0. However, since the number of states increases exponentially as V increases, convergence of energy of $|\psi_1(u_0, k)\rangle$ to u_0 is slow.

(2)

$$|\psi_2(l, k)\rangle \equiv (l - \hat{h})^k |\psi_0\rangle, \tag{4.3}$$

where l is an arbitrary parameter such that $l \geq \max(\hat{h})$. In order to construct $|\psi_2(l, k)\rangle$ which have energy u_0, l, and k should be chosen properly.

(3)

$$|\psi_3(u_0, k)\rangle \equiv \left(\frac{1}{\hat{h} - u_0}\right)^k |\psi_0\rangle. \tag{4.4}$$

Since this resolvent has divergence at $u = u_0$, the degree k is smallest among these three to make the state converge to the energy shell. However, this resolvent is so strong that $|\psi_3\rangle$ becomes a superposition of small number of energy eigenstates whose energies are very close to u_0. The heart of the strong typicality was that the state $|\psi_E\rangle$ is a superposition of exponentially large number of energy eigenstates. Thus, $|\psi_3\rangle$ is not suitable for effective construction of the TPQ state. In addition to this, the resolvent of Hamiltonian is usually harder to calculate than polynomials of Hamiltonian.

Among these three ways to construct the TPQ state, (2) is the most efficient. Hence, I call $|\psi_2(l, k)\rangle$ the "microcanonical TPQ state (mTPQ state)" and write it simply as $|k\rangle$. In the following subsections, I see properties of the mTPQ state.

4.2 Mechanism

In this subsection, I show why the mTPQ state has a sharp-peak energy distribution in the energy shell. The discussion is similar to that in Sect. 3.4, but I obtain the formula to estimate temperature here.

Using Eq. (3.23), the energy distribution $h(u)$ for the microcanonical TPQ state $|k\rangle \equiv (l - \hat{h})^k |\psi_0\rangle$, is

$$h(u) \equiv \sum_{n \in \mathscr{S}(u)} |c_n|^2 (l - u)^{2k} \tag{4.5}$$

$$\simeq g(u)(l - u)^{2k} \tag{4.6}$$

$$= \exp[V\{s(u; V) + 2\kappa \ln(l - u)\}] \tag{4.7}$$

$$= \exp[V\xi_\kappa(u; V)], \tag{4.8}$$

where $\xi_\kappa(u; V) \equiv s(u; V) + 2\kappa \ln(l - u)$. The rhs of Eq. (4.8) has a peak at u_κ^* such that

$$\beta(u;V) = -\frac{\partial\left(2\kappa\ln(l-u)\right)}{\partial u} \tag{4.9}$$

$$= \frac{2\kappa}{(l-u_\kappa^*)} \tag{4.10}$$

where $\beta(u;V) \equiv \frac{\partial s(u;V)}{\partial u}$. I notice that $\beta(u;V)$ is not a parameter like in the cTPQ state but a function derived from entropy of finite V. I expand the rhs of Eq. (4.8) around $u = u^*$. Dropping exponentially small order terms, I get

$$h(u) = \exp\Big[V\Big\{\ \xi_\kappa(u_\kappa^*;V) - \frac{1}{2}|\xi_\kappa''|(u-u_\kappa^*)^2 \\ -\frac{1}{6}\xi_\kappa'''(u-u_\kappa^*)^3 + o\left((u-u_\kappa^*)^4\right)\Big\}\Big]. \tag{4.11}$$

where $\xi_\kappa'' \equiv \xi_\kappa''(u_\kappa^*;V)$ and $\xi_\kappa''' \equiv \xi_\kappa'''(u_\kappa^*;V)$. Since I have assumed $\frac{\partial^2 s(u^*;V)}{\partial u^2} \leq 0$, $\xi_\kappa''(u_\kappa^*;V) = -\Theta(1) \leq 0$. Therefore, $h(u)$ behaves like a Gaussian distribution of the width $\Theta(\frac{1}{\sqrt{V}})$ with the peak at $u = u_\kappa^*$. The energy distribution of Eq. (4.11) is the same as that of a density matrix

$$\rho_k \equiv (l-\hat{h})^{2k}. \tag{4.12}$$

These energy distributions converge to the delta function at $u = u_\kappa^*$ as $N \to \infty$, Hence, I call this ensemble the smooth microcanonical ensemble (because the energy distribution is smooth). Although the mTPQ state is not specified u_κ^* explicitly, it automatically has the sharp energy peak ruled by Eq. (4.10). Therefore, the mTPQ state is definitely the TPQ state at energy u_κ^*.

4.3 Main Results

Using the mTPQ state, all the variables of statistical-mechanical interest are obtained. I show the formulas for temperature, entropy, and mechanical variables.

Equation (4.10) was the condition that the energy distribution has the peak value. I can also interpret this equation as the formula to give temperature. For this purpose, I substitute u_κ^* by

$$u_\kappa \equiv \frac{\langle k|\hat{h}|k\rangle}{\langle k|k\rangle} \tag{4.13}$$

and get

$$\beta(u;V) \sim \frac{2\kappa}{(l-u_\kappa)}. \tag{4.14}$$

To estimate the error of this formula, I calculate the difference between u_κ^* and u_κ. Using Eq. (4.11),

$$u_\kappa - u_\kappa^* = \frac{\int (u - u_\kappa^*) h(u) \mathrm{d}u}{Q_k} \tag{4.15}$$

where $Q_k \equiv \int h(u)\mathrm{d}u$. The numerator is

$$\begin{aligned} &\int e^{V[\xi_\kappa(u_\kappa^*;V) - \frac{1}{2}|\xi_\kappa''|(u-u_\kappa^*)^2]} \left((u - u_\kappa^*) + \frac{V|\xi_\kappa'''|}{6}(u - u_\kappa^*)^4 + O\left((u - u_\kappa^*)^5\right) \right) \mathrm{d}u \\ =& e^{V\xi_\kappa(u_\kappa^*;V)} \left[\frac{\xi'''}{2V\xi''^2} \sqrt{\frac{2\pi}{V|\xi''|}} + O\left(\frac{1}{V^{5/2}}\right) \right] \end{aligned} \tag{4.16}$$

The denominator is

$$\begin{aligned} &\int e^{V[\xi_\kappa(u_\kappa^*;V) - \frac{1}{2}|\xi_\kappa''|(u-u_\kappa^*)^2]} \left(1 - \frac{|\xi_\kappa'''|}{6}(u - u_\kappa^*)^3 + O\left((u - u_\kappa^*)^4\right) \right) \mathrm{d}u \\ =& e^{V\xi_\kappa(u_\kappa^*;V)} \left[\sqrt{\frac{2\pi}{V|\xi_\kappa''|}} + O\left(\frac{1}{V^{3/2}}\right) \right] \end{aligned} \tag{4.17}$$

By using Eqs. (4.16) and (4.17), u_κ^* is close to u_κ as

$$u_\kappa^* = u_\kappa + \frac{\xi'''}{2V\xi''^2} + O\left(\frac{1}{V^2}\right). \tag{4.18}$$

Hence, I can add the N-dependence to Eq. (4.14).

$$\beta(u; V) = \frac{2\kappa}{(l - u_\kappa)} + O\left(\frac{1}{V^1}\right) \tag{4.19}$$

Moreover, using Eq. (4.18), I get a better formula,

$$\beta(u; V) = \frac{2\kappa}{(l - u_\kappa^\bullet)} + O\left(\frac{1}{V^2}\right) \tag{4.20}$$

where

$$u_\kappa^\bullet \equiv u_\kappa + \frac{\xi'''}{2V\xi''^2}. \tag{4.21}$$

$u_\kappa^\bullet$ is $\frac{1}{V}$ order correction for the energy u_κ. I note that one can straightforwardly get the higher order correction terms by evaluating the higher orders of Eqs. (4.16) and (4.17).

Once one calculates temperature, entropy is the integral of it.

$$s(u;V) = \int \beta(u;V)\mathrm{d}u. \tag{4.22}$$

However, one can also calculate entropy directly in the similar manner as Eq. (3.10). By dropping exponentially small order terms, I get

$$\ln\langle k|k\rangle = \ln Q_k. \tag{4.23}$$

From this equation and using Eq. (4.17), I obtain the formula for entropy

$$s(u_\kappa^*;V) = \frac{\ln Q_k}{V} - 2\kappa\ln(l-u_\kappa^*) + \frac{\ln V}{2V} - \frac{1}{2V}\ln\left(\frac{2\pi}{|\xi_\kappa''|}\right) + O\left(\frac{1}{V^2}\right). \tag{4.24}$$

For this equation, I estimate the error caused by $\{z_i\}_i$ in the same way as Eq. (3.14) and get

$$\mathrm{P}\left(\left|\frac{\langle k|k\rangle}{Q_k} - 1\right| \geq \epsilon\right) \leq \frac{1}{\epsilon^2}\frac{Q_{2k}}{Q_k^2}. \tag{4.25}$$

Since it can be proved that $\frac{Q_k^2}{Q_{2k}}$ is $\exp[-\Theta(N)]$, I can say that a single realization of the mTPQ state gives entropy. It may help for understanding to rewrite the rhs by using Eq. (4.24). Then, I get

$$\frac{Q_{2k}}{Q_k^2} = \frac{1}{\exp\left[V\left\{2s(u_{\kappa^*};V) - s(u_{2\kappa^*};V) - 4\kappa\ln\left(\frac{l-u_{2\kappa^*}}{l-u_{\kappa^*}}\right)\right\} + O(\ln V)\right]} \tag{4.26}$$

Very roughly speaking, this result means that the error of the mTPQ state is vanishing in proportional to the number of state, $W(u_{\kappa^*}) \equiv \exp[Vs(u_{\kappa^*};V)]$. It reminds us the result in Sect. 2.2

I can also show that the expectation value of the mTPQ state

$$\langle\hat{A}\rangle_{k,V}^{\mathrm{TPQ}} \equiv \frac{\langle k|\hat{A}|k\rangle}{\langle k|k\rangle} \tag{4.27}$$

is very close to the corresponding ensemble average

$$\langle A\rangle_{k,V}^{\mathrm{ens}} \equiv \frac{\mathrm{Tr}[\hat{A}\rho_k]}{\mathrm{Tr}[\rho_k]}. \tag{4.28}$$

I evaluate the difference between them, and get

$$
\mathrm{P}\left(\left|\langle\hat{A}\rangle_{k,V}^{\mathrm{TPQ}}-\langle A\rangle_{k,V}^{\mathrm{ens}}\right|\geq\epsilon\right)
\leq\frac{1}{\epsilon^{2}}\frac{\langle(\Delta\hat{A})^{2}\rangle_{2k,V}^{\mathrm{ens}}+(\langle A\rangle_{2k,V}^{\mathrm{ens}}-\langle A\rangle_{k,V}^{\mathrm{ens}})^{2}}{\exp\left[V\left\{2s(u_{\kappa^{*}};V)-s(u_{2\kappa^{*}};V)-4\kappa\ln\left(\frac{l-u_{2\kappa^{*}}}{l-u_{\kappa^{*}}}\right)\right\}+O(\ln V)\right]}. \tag{4.29}
$$

where $\langle(\Delta\hat{A})^{2}\rangle_{k,V}^{\mathrm{ens}}\equiv\langle(\hat{A}-\langle\hat{A}\rangle_{k,V}^{\mathrm{ens}})^{2}\rangle_{k,V}^{\mathrm{ens}}$. Since the denominator is $\exp[\Theta(V)]$ and the numerator is $O(V^{2m})$, the difference between $\langle\hat{A}\rangle_{k,V}^{\mathrm{TPQ}}$ and $\langle A\rangle_{k,V}^{\mathrm{ens}}$ is $\exp[-\Theta(V)]$.

4.4 Comparison Between cTPQ State and mTPQ State

In this section, I introduced the mTPQ state to construct the TPQ state easily in practical applications. Actually, the mTPQ state can be generated just by multiplying the Hamiltonian matrix to the initial random vector. However, the microcanonical ensemble has the ambiguity such that the width of the energy shell is not unique but is tolerated as far as it vanishes in the thermodynamic limit, i.e., $o(V)$. As a consequence, the results of the mTPQ state have the correction terms. For example, in Eq. (4.24), Q_k is not directly related to the entropy density, $s(u_\kappa^*, V)$, but the equation has many correction terms. When I compare these merits and demerits to the main results of the cTPQ state, the latter is concise with no correction terms, but $\exp[-\frac{1}{2}\beta\hat{h}]$ is difficult to calculate in most cases. To combine the merit of the mTPQ state and that of the cTPQ state, I will show in Chap. 6 that the cTPQ state is effectively generated from the mTPQ states.

I note that the mTPQ state is advantageous when system is at a first-order phase transition [2]. In general,

$$
\frac{\partial}{\partial u}\beta(u;V)\leq 0 \tag{4.30}
$$

holds asymptotically, where the equality holds only at the first-order phase transition. Hence,

$$
\frac{\partial^{2}}{\partial u^{2}}\xi_{\kappa}(u;V)=\frac{\partial}{\partial u}\beta(u;V)-\frac{2\kappa}{(l-u)^{2}}<0 \tag{4.31}
$$

holds for every finite u, even at a first-order phase transition. As a result, Eq. (4.17) is valid even when a first-order phase transition takes place. (This is an advantage of our introducing l!)

That is, at a first-order phase transition $\beta(u_\kappa^*; N)$ takes the same value for multiple values of κ (and u_κ^*). For each of such κ, however, Eq. (4.17) is valid. and the solution u_κ^* of Eq. (4.10) is determined uniquely.

At a first-order phase transition an equilibrium state cannot be specified by (β, N) uniquely. [See Ref. [1] for complete discussions, and Ref. [2] for a hint.] To specify an equilibrium state uniquely, one must use (u, N) instead of (β, N). As a result of this fact, neither the canonical density operator (of the ensemble formulation) nor the canonical TPQ state (of our formulation) can specify an equilibrium state uniquely. One must use either the microcanonical density operator (of the ensemble formulation) nor the microcanonical TPQ state (of our formulation) to specify an equilibrium state uniquely, at a first-order phase transition.

Nevertheless, one can use the canonical density operator (of the ensemble formulation) or the canonical TPQ state (of our formulation) even at a first-order phase transition, because they give the correct free energy, from which the entropy function (the fundamental relation) can be obtained by the Legendre transformation [1]. As a result, all the formulas of our paper are valid even at a first-order phase transition.

References

1. A. Shimizu, Netsurikigaku no Kiso (Principles of Thermodynamics) (University of Tokyo Press, Tokyo, 2007) [in Japanese].
2. A. Shimizu, J. Phys. Soc. Jpn. **77**, 104001 (2008)

Chapter 5
Equilibrium State and Entanglement

Up to here, I have explained how to produce statistical mechanics based on a pure quantum state. Conventionally, some quantum mixed states such as the Gibbs state may be regarded as the only equilibrium state. However, I have showed that there is a wide range of possibilities that many states including a pure quantum state can represent the equilibrium state. In this section, I discuss the similarity and the difference between the TPQ formulation and the ensemble one.

I firstly investigate what is implied when a state is regarded as the equilibrium state. I show that the TPQ state is macroscopically time-independent and cannot be distinguished from the conventional equilibrium states. I also show that the TPQ state can predict the linear response theory.

Second, I discuss fluctuation [1]. I reveal that a distinction between so-called thermal fluctuation and quantum fluctuation is impossible in mixed quantum states. I also discuss that the TPQ state does not have any thermal fluctuation but all the fluctuation is squeezed into the quantum one.

Third, I show that quantum entanglement is one of the ways to detect the difference between the TPQ states and the conventional states [1]. It directly leads to a stability of pure quantum states. It turns out that the TPQ state is stable state against a weak classical noise.

5.1 Time Evolution

To begin with, let us consider the time evolution of the Gibbs state ρ_β. Since ρ_β commutes with Hamiltonian, it is time-invariant:

$$e^{\frac{i}{\hbar}\hat{H}}e^{-\beta\hat{H}}e^{-\frac{i}{\hbar}\hat{H}} = e^{-\beta\hat{H}} \tag{5.1}$$

S. Sugiura, *Formulation of Statistical Mechanics Based on Thermal Pure Quantum States*, Springer Theses, DOI 10.1007/978-981-10-1506-9_5

On the other hand, the time evolution changes the TPQ state:

$$|\beta, V(t)\rangle \equiv e^{-\frac{i}{\hbar}\hat{H}t}|\beta, V\rangle \tag{5.2}$$

$$= \sum_n z_n e^{-\frac{i}{\hbar}u_n t} e^{-\frac{1}{2}\beta u_n}|n\rangle \tag{5.3}$$

$$\neq |\beta, V\rangle. \tag{5.4}$$

However, the random variables $\{z_n\}_n$ have an invariant probability measure under any unitary transformation. Hence, all the result in Chap. 3 still holds for $|\beta, V(t)\rangle$, and it is regarded as another realization of the cTPQ state. Namely, although the cTPQ state microscopically changes over time, it stays macroscopically invariant. This fact closely related to the thermalization [2, 3], because it suggests that a pure quantum state rarely returns to a non-equilibrium state after it evolves into an equilibrium state.

5.2 Linear Response Theory

When a weak probe field $f(t)$ is applied to an equilibrium state, the change of a mechanical variable $\hat{B}$ is predicted by linear response theory. I write time-dependent Hamiltonian as $\hat{H}(t) \equiv \hat{H}_0 - \hat{A}f(t)$, where $\hat{H}_0$ is initial Hamiltonian and $-\hat{A}f(t)$ is a perturbation which consists of the external force $f(t)$ and the corresponding mechanical variable $\hat{A}$. Then, the response against a probe field $f(t)$ is evaluated to a linear order of $f(t)$ as

$$\Delta A(t) = \int_{-\infty}^{t_{\rm fin}} dt' \Phi_{AB}(t-t') f(t'), \tag{5.5}$$

where $\Delta A(t) \equiv \mathrm{Tr}[\hat{\rho}(t)\hat{A}] - \mathrm{Tr}[\hat{\rho}_0\hat{A}]$, $\hat{\rho}(t)$ is a state at $t = t_{\rm fin}$, and $\Phi_{AB}(t-t')$ is a response function.

The Green-Kubo relations give the response function from a time correlation function:

$$\Phi_{AB}(t-t') = \frac{1}{i\hbar}\mathrm{Tr}\left(\rho_0[\hat{B}(t'-t), \hat{A}]\right) \tag{5.6}$$

where $\hat{\rho}_0$ is an initial equilibrium state $\hat{\rho}_0 \equiv \exp[-\beta\hat{H}_0]/Z(\beta, V)$, and $\hat{B}(t'-t) \equiv \exp[\frac{i}{\hbar}\hat{H}_0(t-t')]\hat{B}\exp[-\frac{i}{\hbar}\hat{H}_0(t-t')]$. Hence, the response against some external force is obtained from the corresponding time correlation function of the equilibrium state.

In the TPQ formulation, I substitute ρ_0 with $|\beta, V\rangle$ in Eq. (5.6) and get a relation

$$\Phi_{AB}(t-t') = \frac{1}{i\hbar} \frac{\langle \beta, V | [\hat{B}(t'-t), \hat{A}] | \beta, V \rangle}{\langle \beta, V | \beta, V \rangle}. \tag{5.7}$$

This equation is correct within an exponentially small error. To estimate the error, I substitute $[\hat{B}(t'-t), \hat{A}]$ for $\hat{A}$ in Eq. (3.18) and get

$$\begin{aligned} &\mathrm{P}\left(\left| \langle [\hat{B}(t'-t), \hat{A}] \rangle^{\mathrm{TPQ}}_{\beta,V} - \langle [\hat{B}(t'-t), \hat{A}] \rangle^{\mathrm{ens}}_{\beta,V} \right| \geq \epsilon \right) \\ &\leq \frac{1}{\epsilon^2} \frac{\langle [\hat{B}(t'-t), \hat{A}]^2 \rangle^{\mathrm{ens}}_{2\beta,V} + (\langle [\hat{B}(t'-t), \hat{A}]^2 \rangle^{\mathrm{ens}}_{2\beta,V} - \langle [\hat{B}(t'-t), \hat{A}]^2 \rangle^{\mathrm{ens}}_{\beta,V})^2}{\exp[2V\beta\{f(T/2, \mu; V) - f(T, ; V)\}]}. \end{aligned} \tag{5.8}$$

Since $\| \exp[-\frac{i}{\hbar}\hat{H}t] \| = 1$, the rhs of the numerator is $O(V^{2m})$. Therefore, the rhs is still $\exp[-\Theta(V)]$. Namely, one can evaluate the linear response function using the TPQ state, although the transportation is non-equilibrium phenomena. This is because the linear response theory focuses on mechanical variables. The typicality holds when the number of the physical quantities which we observe is negligibly smaller than the degree of the freedom of the Hilbert space. Therefore, we obtain the correct results almost surely as far as we look at the small number of linear response functions.

5.3 Quantum and Thermal Fluctuations

To better understand the TPQ states, I now discuss the "quantum fluctuation" and "thermal fluctuation." For concreteness, I consider the canonical TPQ state $|\beta, N\rangle$ and the canonical density operator $\hat{\rho}_\beta$.

In the ensemble formulation, it is often said that a fluctuation of a mechanical variable $\langle (\Delta\hat{A})^2 \rangle^{\mathrm{ens}} \equiv \langle (\hat{A} - \langle \hat{A} \rangle^{\mathrm{ens}})^2 \rangle^{\mathrm{ens}}$ can be decomposed into the quantum fluctuation $\langle (\Delta\hat{A})^2 \rangle^{\mathrm{ens}}_{\mathrm{q}}$ and the thermal one $\langle (\Delta\hat{A})^2 \rangle^{\mathrm{ens}}_{\mathrm{t}}$, i.e.,

$$\langle (\Delta\hat{A})^2 \rangle^{\mathrm{ens}} = \langle (\Delta\hat{A})^2 \rangle^{\mathrm{ens}}_{\mathrm{q}} + \langle (\Delta\hat{A})^2 \rangle^{\mathrm{ens}}_{\mathrm{t}}. \tag{5.9}$$

The thermal fluctuation, whose specific expression will be given below, is conventionally interpreted as a result of mixing many quantum states to form $\hat{\rho}$,

$$\hat{\rho}_\beta = \sum_n (e^{-\beta N u_n}/Z) |n\rangle\langle n|, \tag{5.10}$$

where u_n and $|n\rangle$ are eigenvalue and eigenstate, respectively, of $\hat{h}$. Consequently, it is conventionally concluded that the thermal fluctuation of most mechanical variables does not vanish at any finite temperature.

In the TPQ formulation, by contrast, $|\beta, V\rangle$ is a pure quantum state and therefore does not have such "thermal fluctuation," i.e., $\langle (\Delta\hat{A})^2 \rangle^{\mathrm{TPQ}}_{\mathrm{t}} = 0$ at all temperature.

The TPQ state has only the quantum fluctuation, i.e.,

$$\langle(\Delta\hat{A})^2\rangle^{\mathrm{TPQ}} = \langle(\Delta\hat{A})^2\rangle^{\mathrm{TPQ}}_{\mathrm{q}} \equiv \langle(\hat{A} - \langle\hat{A}\rangle^{\mathrm{TPQ}})^2\rangle^{\mathrm{TPQ}}. \tag{5.11}$$

In other words, all fluctuations are included in the quantum fluctuation.

I have thus found that $\hat{\rho}_\beta$ and $|\beta, V\rangle$, which represent the same equilibrium state and give different values of the quantum and thermal fluctuations. This does not lead to any contradiction in experimentally observable quantities because

$$\langle(\Delta\hat{A})^2\rangle^{\mathrm{ens}} = \langle(\Delta\hat{A})^2\rangle^{\mathrm{TPQ}}, \tag{5.12}$$

which are the only observable quantities in the above discussion. The quantum and thermal fluctuations, $\langle(\Delta\hat{A})^2\rangle^{\mathrm{ens}}_{\mathrm{q}}$ and $\langle(\Delta\hat{A})^2\rangle^{\mathrm{ens}}_{\mathrm{t}}$, are, separately, not observable quantities. To see this, let us write them down explicitly. I note that ρ has the following form,

$$\hat{\rho} \equiv \sum_\lambda w_\lambda |\lambda\rangle\langle\lambda|, \tag{5.13}$$

where $\{w_\lambda\}_\lambda$ is a set of positive numbers such that $\sum_\lambda w_\lambda = 1$, and $\{|\lambda\rangle\}_\lambda$ is some set of states (which is $\{|n\rangle\}_n$ in Eq. (5.10)). In general, $\hat{A}$ fluctuates quantum mechanically in each state $|\lambda\rangle$. Hence, it may be reasonable to define $\langle(\Delta\hat{A})^2\rangle^{\mathrm{ens}}_{\mathrm{q}}$ as the average of the fluctuation $\langle\lambda|(\hat{A} - \langle\lambda|\hat{A}|\lambda\rangle)^2|\lambda\rangle$ over $|\lambda\rangle$'s, i.e.,

$$\langle(\Delta\hat{A})^2\rangle^{\mathrm{ens}}_{\mathrm{q}} \equiv \sum_\lambda w_\lambda \langle\lambda|(\hat{A} - \langle\lambda|\hat{A}|\lambda\rangle)^2|\lambda\rangle. \tag{5.14}$$

This and Eq. (5.9) yield the thermal fluctuation as

$$\langle(\Delta\hat{A})^2\rangle^{\mathrm{ens}}_{\mathrm{t}} = \sum_\lambda w_\lambda \langle\lambda|\hat{A}|\lambda\rangle^2 - \left(\sum_\lambda w_\lambda \langle\lambda|\hat{A}|\lambda\rangle\right)^2. \tag{5.15}$$

If I take $w_\lambda = e^{-\beta N u_n}/Z$ and $|\lambda\rangle = |n\rangle$, I find that $\langle(\Delta\hat{A})^2\rangle^{\mathrm{ens}}_{\mathrm{t}} > 0$ for most mechanical variables at finite temperature.

However, it is well-known that $|\lambda\rangle$'s in Eq. (5.13) need not be orthogonal to each other [4]. As a result, there are infinitely many possible choices of $\{|\lambda\rangle\}_\lambda$ and $\{w_\lambda\}_\lambda$ for the same $\hat{\rho}$ [4]. The experimentally observable fluctuation $\langle(\Delta\hat{A})^2\rangle^{\mathrm{ens}}$ is invariant under the change of $\{w_\lambda\}_\lambda$ and $\{|\lambda\rangle\}_\lambda$. By contrast, both $\langle(\Delta\hat{A})^2\rangle^{\mathrm{ens}}_{\mathrm{q}}$ and $\langle(\Delta\hat{A})^2\rangle^{\mathrm{ens}}_{\mathrm{t}}$ do alter under the change of $\{w_\lambda\}_\lambda$ and $\{|\lambda\rangle\}_\lambda$. This fact clearly shows that the quantum and thermal fluctuations are, separately, not experimentally observable quantities. In other words, they are, separately, metaphysical quantities.

It is instructive to consider a classical mixture

$$\hat{\rho}'_\beta \equiv \frac{1}{R}\sum_{r=1}^{R} \frac{|\beta, V, r\rangle\langle\beta, V, r|}{\langle\beta, V, r|\beta, V, r\rangle} \tag{5.16}$$

of many realizations $|\beta, V, 1\rangle, |\beta, V, 2\rangle, \cdots, |\beta, V, R\rangle$ of the canonical TPQ state. Since each $|\beta, V, r\rangle$ represents the same equilibrium state, so does $\hat{\rho}'_\beta$. If I define the quantum and thermal fluctuations in $\hat{\rho}'_\beta$ in the same way as Eqs. (5.14) and (5.15), I find that the thermal fluctuation is exponentially small for all mechanical variables. This shows that mixing many states does not necessarily give "thermal fluctuation." Since the thermal fluctuation in $\hat{\rho}'_\beta$ is negligible, I do not need to take an average over many realizations, but only need to pick up a single realization.

5.4 Entanglement

In macroscopic view point, the ensemble formulation and the TPQ formulation give identical results for all quantities of statistical-mechanical interest. That is, as far as one looks at macroscopic quantities, one cannot distinguish between these states. However, when I look at the quantum entanglement, they are completely different.

5.4.1 Entanglement Measure for Pure State

When I measure the bipartite entanglement of a state $|\psi\rangle$ on a system, I divide the system into two parts. Let us call them a system A and a system B. Then, I trace out the components of $|\psi\rangle$ on the system B.

$$\rho_A \equiv \mathrm{Tr}_B[|\psi\rangle\langle\psi|] \tag{5.17}$$

The von Neumann entropy is defined as

$$E_{vN}(|\psi\rangle\langle\psi|) \equiv \mathrm{Tr}[-\rho_A \ln \rho_A], \tag{5.18}$$

and the purity is defined as

$$E_{purity}(|\psi\rangle\langle\psi|) \equiv \mathrm{Tr}[\rho_A^2]. \tag{5.19}$$

When the state $|\psi\rangle$ is pure, they can measure how much quantum entanglement exists between A and B. As a state gets more entanglement between A and B, E_{vN} takes larger value and E_{purity} takes smaller value. In particular, when there is no entanglement between them, $E_{vN} = 0$ and $E_{purity} = 1$. Since ρ_A and ρ_B have the same entanglement, I assume that the size of the system A is smaller than that of the system B without loss of generality.

5.4.2 Entanglement Measure for Mixed State

When the state is mixed, classical mixture reduces the quantum entanglement. A mixed state is in general represented as

$$\hat{\rho} \equiv \sum_{i,j} w_{i,j} |i\rangle\langle j| \tag{5.20}$$

where $w_{i,j}$ is a positive definite matrix and $\sum_i w_{i,i} = 1$, and $\{|i\rangle\}_i$ is a set of orthonormal basis. By diagonalizing the matrix $w_{i,j}$, I get

$$\hat{\rho} = \sum_{n} w_n |n\rangle\langle n| \tag{5.21}$$

where w_n is a probability such that $\sum_n w_n = 1$, and $\{|n\rangle\}_n$ is a set of basis which can be non-orthogonal. Since $\{|n\rangle\}_n$ is tolerated to be non-orthogonal, the set of basis $\{|n\rangle\}_n$ is not always unique. I define a group of the tolerated basis sets as

$$\mathscr{D}_{\hat{\rho}} \equiv \left\{ \{|n\rangle\}_n \middle| \{|n\rangle\}_n \text{ diagonalizes } w_{i,j} \right\} \tag{5.22}$$

Since there are many representations in the form of Eq. (5.21), I cannot use either Eq. (5.18) or (5.19) directly but need to define a new probability measure. Although there are many definitions of the entangle measure for the mixed states, I introduce "entanglement of formation" here. The entanglement of formation E_{form} is defined by

$$E_{\text{form}}(\hat{\rho}) \equiv \min_{\{|n\rangle\}_n \in \mathscr{D}_{\hat{\rho}}} \left\{ \sum_n w_n E_{\text{vN}} \left(|n\rangle\langle n| \right) \right\} \tag{5.23}$$

I note that $E_{\text{form}}(\hat{\rho})$ reduces to $E_{\text{vN}}(\hat{\rho})$ when $\hat{\rho}$ is a pure state.

5.4.3 Entanglement at Finite Temperature in Ensemble Formulation

When I employ the canonical ensemble in the ensemble formulation, the equilibrium state is represented by the Gibbs state $\hat{\rho}_\beta$. I consider the quantum entanglement of it. However, it is known that the calculation of the quantum entanglement for the mixed state is very hard (and NP-hard in many cases [5]). Thus, I consider $\beta = 0$ case and $\beta \to \infty$ case.

In the former case, the Gibbs state reduces to an identical operator

$$\frac{\hat{1}}{D} \equiv \frac{1}{D}\sum_i |i\rangle\langle i| \tag{5.24}$$

where D is the dimension of Hilbert space, and $\{|i\rangle\}_i$ is an arbitrary orthonormal basis set. In the case of $\beta = 0$, when I use a set of product states as $\{|i\rangle\}_i$ in Eq. (5.24), for ${}^\forall|i\rangle$,

$$E_{\mathrm{vN}}\left(|i\rangle\langle i|\right) = 0. \tag{5.25}$$

Therefore, by using Eq. (5.23), I get

$$E_{\mathrm{form}}\left(\frac{\hat{1}}{D}\right) = 0. \tag{5.26}$$

Namely, $\hat{\rho}_\beta$ has no entanglement at $\beta = 0$.

In the case of $\beta \to \infty$, $\exp[-\beta\hat{H}]$ reduces to a ground state. I consider the case that symmetry of the state is broken and there is no degeneracy in the ground state. Thus, $\hat{\rho}_\beta$ reduces to a pure quantum ground state $|\psi\rangle\langle\psi|$. Thus, the quantum entanglement is

$$E_{\mathrm{form}}\left(|\psi\rangle\langle\psi|\right) = E_{\mathrm{vN}}\left(|\psi\rangle\langle\psi|\right). \tag{5.27}$$

In general, the value of $E_{\mathrm{vN}}\left(|\psi\rangle\langle\psi|\right)$ is nonzero except for some trivial models.

5.4.4 Entanglement at Finite Temperature in TPQ Formulation—Analytics

Since in the TPQ formulation, the equilibrium states are represented by the pure quantum states, the quantum entanglement is well-defined and easy to measure. The result is very different from that of the ensemble formulation. I analyze the quantum entanglement of the (normalized) cTPQ state

$$|\psi_\beta\rangle \equiv \frac{|\beta, V\rangle}{\sqrt{\langle\beta, V|\beta, V\rangle}} \tag{5.28}$$

for the four cases: (i) $\beta = 0$, (ii) $\beta \to \infty$, (iii) A≪B and A is sufficiently large, and (iv) A and B are sufficiently large.

(i) $\beta = 0$

The result is the same as Eq. (5.27).

(ii) $\beta \to \infty$

According to Ref. [6],

$$\overline{E_{\text{purity}}(|\psi_\beta\rangle\langle\psi_\beta|)} = \overline{\mathrm{Tr}[\hat{\rho}_{\mathrm{A}}^2]} \tag{5.29}$$

$$= \frac{d_{\mathrm{A}} + d_{\mathrm{B}}}{d_{\mathrm{A}} d_{\mathrm{B}} + 1} \tag{5.30}$$

$$= \frac{1}{d_{\mathrm{A}}}\left(1 + \frac{1}{d^{N_{\mathrm{B}} - N_{\mathrm{A}}}} + \cdots\right), \tag{5.31}$$

where $\overline{\cdots}$ denotes average over $\{z_i\}_i$, d is the dimension per site, e.g., $d = 2$ for a spin-1/2 system, N_{A} (N_{B}) is the number of sites in A (B), and the dimension of Hilbert space of subsystem A (B) is $d_{\mathrm{A}} \equiv d^{N_{\mathrm{A}}}$ ($d_{\mathrm{B}} \equiv d^{N_{\mathrm{B}}}$). The minimum value of the purity for the number of sites being N_{A} is $\frac{1}{d_{\mathrm{A}}}$, which is achieved in the case of $\hat{\rho_{\mathrm{A}}} = \frac{\hat{1}}{d_{\mathrm{A}}}$. Since the rhs of Eq. (5.31) is very close to this minimum value, Eq. (5.31) indicates that the TPQ state for $\beta \to \infty$ has almost maximum entanglement.

(iii) A$\ll$B and A is sufficiently large

In Sect. 2.2.4, I have shown that ρ_{A} becomes close to the Gibbs state of the subsystem A when A$\ll$B is satisfied. That is,

$$\rho_{\mathrm{A}} \simeq \frac{\exp[-\beta \hat{H}_{\mathrm{A}}]}{Z_{\mathrm{A}}(\beta)}, \tag{5.32}$$

where $\hat{H}_{\mathrm{A}}$ is Hamiltonian of the system A, and $Z_{\mathrm{A}}(\beta) \equiv \mathrm{Tr}\left[\exp[-\beta \hat{H}_{\mathrm{A}}]\right]$. Here, I ignore an interaction between A and B, because A is sufficiently large. In this case,

$$E_{\mathrm{vN}}(|\psi_\beta\rangle\langle\psi_\beta|) \simeq S(\beta, N_{\mathrm{A}}) \tag{5.33}$$

$$\equiv \beta\left(\mathrm{Tr}\left[\hat{H}\rho_{\mathrm{A}}\right] - F\left(\frac{1}{\beta}, N_{\mathrm{A}}\right)\right), \tag{5.34}$$

where $F_{\mathrm{A}}(\frac{1}{\beta}, N_{\mathrm{A}}) \equiv \frac{1}{\beta} \ln Z_A(\beta, V)$. Since $S(\beta, N_{\mathrm{A}})$ is $\Theta(N_{\mathrm{A}})$ at finite temperature, this result means that the cTPQ state is highly entangled. For the purity, I get

$$E_{\text{purity}}(|\psi_\beta\rangle\langle\psi_\beta|) \simeq \frac{Z_A(2\beta, V)}{Z_A(\beta, V)^2} \tag{5.35}$$

$$= \frac{1}{\exp[2\beta(F_{\mathrm{A}}(\frac{1}{2\beta}, N_{\mathrm{A}}) - F_{\mathrm{A}}(\frac{1}{\beta}, N_{\mathrm{A}}))]}. \tag{5.36}$$

(The rhs is mostly the same as the rhs of Eq. (3.12).) Since the rhs of Eq. (5.36) is $\exp[-\Theta(N_{\mathrm{A}})]$, the purity also indicates that the cTPQ state is highly entangled.

(iv) A and B are sufficiently large

A logarithm of the purity is called second Renyi entropy:

$$E_2 \equiv -\ln[\mathrm{Tr}(\hat{\rho}_q^2)]. \tag{5.37}$$

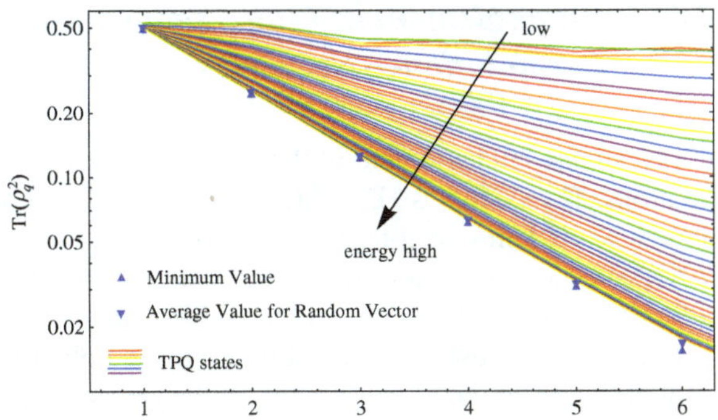

Fig. 5.1 Purity versus q of the 1D Heisenberg chain for $N = 16$

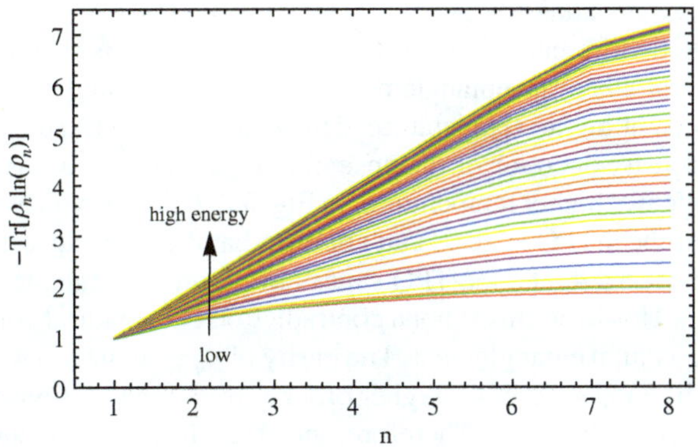

Fig. 5.2 Purity versus q of the 1D Heisenberg chain for $N = 16$

As we see in Fig. 5.1, the second Renyi entropy of the TPQ state increases linearly as q increases. However, as q further increases, it starts to deviate from the linear function around $q \sim N/2$ and decreases in $q > N/2$. This complete size dependence of the second Renyi entropy of the cTPQ state can be calculated analytically. According to Ref. [7], it becomes

$$E_2(|\psi_\beta\rangle\langle\psi_\beta|) \simeq -\ln\left[\frac{1}{\exp[2\beta(F_\mathrm{A}(\frac{1}{2\beta}, N_\mathrm{A}) - F_\mathrm{A}(\frac{1}{\beta}, N_\mathrm{A}))]} + \frac{1}{\exp[2\beta(F_\mathrm{B}(\frac{1}{2\beta}, N_\mathrm{B}) - F_\mathrm{B}(\frac{1}{\beta}, N_\mathrm{B}))]}\right], \quad (5.38)$$

where $\hat{H}_\mathrm{B}$ is Hamiltonian of the system B, and $F(\frac{1}{\beta, N_B}) \equiv \frac{1}{\beta} \ln \mathrm{Tr}\left[\exp[-\beta \hat{H}_\mathrm{B}]\right]$. Eq. (5.38) is symmetric when one exchanges A and B. This structure of the entanglement entropy is called a "Page curve" after a seminal work by D. Page [8]. In Sect. 5.4.5, we see that this structure appears in results of numerical calculation. Furthermore, quantitative analysis is done in Ref. [7], and it is shown that Eq. (5.38) explains not only the second Renyi entropy of cTPQ states but also those of a broad class of pure quantum states.

5.4.5 *Entanglement at Finite Temperature in TPQ Formulation—Numerics*

To investigate entanglement of the TPQ state further, I perform some numerical calculations. The system is 1D antiferromagnetic Heisenberg model with the number of sites $N = 16$. I choose consecutive q sites as the system A, and the other consecutive $N - q$ sites are the system B. I prepare the mTPQ state for different energies. Their reduced density operator ρ_q is obtained by tracing out the system B.

In Fig. 5.1, I show the result of their purity. I plot the minimum value of the purity (triangles $\triangle$) and the average value of the purity of the random vector $|\psi_0\rangle$ (inverse triangles $\triangledown$), which appeared in Eq. (5.31). It is seen that $|\psi_0\rangle$ has almost maximum (exponentially large) entanglement [6]. The lines are the purity of the mTPQ states with different values of the energy density. It is seen that the TPQ states have exponentially large entanglement, and that the entanglement gets larger at higher energy, i.e., at higher temperature. This result is in marked contrast to entanglement of the density operator of the ensemble formulation, because the latter has less entanglement at higher temperature. In Fig. 5.2, I plot the von Neumann entanglement entropy of the mTPQ states. I can confirm that these results are consistent with the results in Sect. 5.4.4, and the TPQ states have large entanglement.

However, this is not a contradiction but a natural consequence of the nature of the quantum entanglement. The purity of ρ_q is related to N-body correlation functions of the TPQ state. Such higher order correlation functions represent microscopic details of the TPQ state. Therefore, the great difference in entanglement between the TPQ states and the Gibbs states indicates a great difference in microscopic details. It is not surprising that such *microscopically* completely different states give identical results for *macroscopic* quantities and thus represent the same equilibrium state.

5.5 Stability Against Noise

The TPQ states have the large quantum entanglement. In quantum information, one may think that a state with large quantum entanglement is unstable against noise, and the state will decohere immediately. In fact, it is true for Schoedinger cat states. However, I show in this subsection that the quantum entanglement of the TPQ states is resistant against a weak classical noise.

The Schoedinger cat state is represented by the superposition of macroscopically distinct states. For example,

$$|\psi_{\text{cat}}\rangle = \frac{1}{2}(|\downarrow\downarrow\downarrow\cdots\downarrow\rangle + |\uparrow\uparrow\uparrow\cdots\uparrow\rangle) \tag{5.39}$$

is the superposition of the state with $S_z = \frac{N\hbar}{2}$ and that with $S_z = -\frac{N\hbar}{2}$ where N is the number of spins. Thus, the fluctuation of S_z is macroscopically large;

$$\frac{\sqrt{\langle\psi_{\rm cat}|\hat{S}_z^2|\psi_{\rm cat}\rangle - \langle\psi_{\rm cat}|\hat{S}_z|\psi_{\rm cat}\rangle^2} = N\hbar}{2,} \tag{5.40}$$

which increases linearly as N increases. By contrast, a normal quantum state in a usual experimental condition has a small quantum fluctuation. For example,

$$|\psi_{\rm real}\rangle = \frac{1}{N}(|\downarrow\uparrow\uparrow\cdots\uparrow\rangle + |\uparrow\downarrow\uparrow\cdots\uparrow\rangle + |\uparrow\uparrow\downarrow\cdots\uparrow\rangle + |\uparrow\uparrow\uparrow\cdots\downarrow\rangle) \tag{5.41}$$

is the superposition of N states. However, these states are macroscopically the same states because all the states have $S_z = \frac{(N-1)\hbar}{2}$. Hence, the fluctuation of S_z is

$$\langle\psi_{\rm real}|\hat{S}_z^2|\psi_{\rm real}\rangle - \langle\psi_{\rm real}|\hat{S}_z|\psi_{\rm real}\rangle^2 = 0. \tag{5.42}$$

(Strictly speaking, the state with no fluctuation may be also unrealistic. However, such a state can be realized as a ground state.)

In order to detect the difference between the cat states and the normal states, I use the index "p," which is introduced by Shimizu and Miyadera [9]. Let $\hat{A}$ be a mechanical variable with the degree $m = 1$. I call such a variable an "additive operator." I evaluate the maximum value of the fluctuation among $\hat{A}$.

$$\max_{\hat{A}\in {\rm additiveoperator}} \left(\langle\psi|\hat{A}^2|\psi\rangle - \langle\psi|\hat{A}|\psi\rangle^2\right) = O(V^p). \tag{5.43}$$

Here, I define the index "p." Using this index, I can characterize the normal states as $p = 1$ and the cat states as $p = 2$. Then, I will show that the cat states are fragile against some weak classical noises.

For a system with a weak classical noise, the locality requires that the Hamiltonian of the noise is sum of local interactions,

$$\hat{H}_{\rm int} = \lambda \sum_x f(x,t)\hat{a}(x). \tag{5.44}$$

Here, λ is a positive small constant, x is an index of a position, $f(x,t)$ is an amplitude of the noise with vanishing average over time, and $\hat{a}(x)$ is a local operator at x. When a quantum state sustains this weak classical noise, the rate of decoherence is measured by

$$\Gamma \equiv -\frac{1}{2}\frac{\rm d}{{\rm d}t}\left(\ln {\rm Tr}[\hat{\rho}(t)^2]\right)\Big|_{t\ll 1/\Gamma} \tag{5.45}$$

According to Ref. [9], when the p-index for a state is $p = 1$, $\Gamma = O(V)$ for any weak classical noises; when the p-index for a state is $p = 2$, $\Gamma = O(V^{1+\delta})$ $(0 < \delta \le 1)$ for some weak classical noises. The former means that any state with $p = 1$ is stable against any weak classical noises. The latter means that for every cat state there exist some noises which break the quantum entanglement of the state.

Thermodynamics requires that any fluctuations of the additive operators are $\langle(\Delta\hat{A})^2\rangle^{\text{ens}} = O(V)$. Therefore, returning to the TPQ states,

$$\langle(\Delta\hat{A})^2\rangle^{\text{TPQ}} = O(V) \tag{5.46}$$

holds for any additive operators and $p = 1$ for the TPQ states. Therefore, I can say that the TPQ states are stable against any weak classical noise in the sense of Eq. (5.45).

References

1. S. Sugiura, A. Shimizu, Kinki University Series on Quantum Computing **9**, 245 (2014)
2. J. von Neumann, Zeitschrift fuer Physik **57**, 30 (1929) (in German); J. von Neumann, Eur. Phys. J. H **35**, 173 (2010) (English Translation)
3. P. Reimann, Phys. Rev. Lett. **101**, 190403 (2008)
4. M.A. Nielsen, I.L. Chuang, *Quantum Computation and Quantum Information* (Cambridge University Press, Cambridge, 2000)
5. S. Gharibian, Quantum Information and Computation **10**(3&4), 343 (2010)
6. A. Sugita, A. Shimizu, J. Phys. Soc. Jpn. **74**, 1883 (2005)
7. H. Fujita, Y.O. Nakagawa, S. Sugiura, M. Watanabe, arXiv:1703.02993
8. D.N. Page, Phys. Rev. Lett. **71**, 1291 (1993)
9. A. Shimizu, T. Miyadera, Phys. Rev. Lett. **89**, 270403 (2002)

Chapter 6
Relation Among TPQ States

In thermodynamics, any one of the thermodynamic functions such as entropy and free energy is transformed into another one by Legendre transformation.

$$F = E - TS \tag{6.1}$$

$$J = F - \mu N \tag{6.2}$$

Similarly, the TPQ state itself can be transformed into another one [1]. Although the cTPQ states and the gTPQ states have an exponential function of Hamiltonian, which are hard to calculate, this transformation enables us to construct these TPQ states from the mTPQ states, which are easy to construct.

6.1 Decomposition of cTPQ State and gTPQ States

The cTPQ state is decomposed into the superposition of the mTPQ states. I employ the Taylor series expansion.

$$e^{\frac{V\beta l}{2}}|\beta, V\rangle = \sum_{0}^{\infty} \frac{1}{k!}\left(\frac{V\beta}{2}\right)^{k} (l - \hat{h})^{k}|\psi_0\rangle \tag{6.3}$$

$$= \sum_{0}^{\infty} \frac{1}{k!}\left(\frac{V\beta}{2}\right)^{k} |k\rangle \tag{6.4}$$

$$= \sum_{0}^{\infty} R_k|\psi_k\rangle \tag{6.5}$$

S. Sugiura, *Formulation of Statistical Mechanics Based on Thermal Pure Quantum States*, Springer Theses, DOI 10.1007/978-981-10-1506-9_6

where $|\psi_k\rangle$ is a normalized mTPQ state, $|\psi_k\rangle \equiv \frac{1}{\sqrt{Q_k}}|k\rangle$, and $R_k \equiv \frac{1}{k!}\left(\frac{V\beta}{2}\right)^k \sqrt{Q_k}$. Then, I see the convergence of R_k. Using Eq. (4.17) and Stirling's formula, I have

$$R_k \sim \exp\left[\frac{V}{2}\zeta_\kappa(\beta; V)\right], \tag{6.6}$$

where

$$\zeta_\kappa(\beta; V) \equiv \xi_\kappa(u_\kappa^*; V) + 2\kappa \ln(\beta/2\kappa) + 2\kappa. \tag{6.7}$$

Here, β in the rhs is not a function $\beta(u_\kappa^*; N)$ but an independent parameter of the canonical TPQ state. If κ took continuous values, ζ_κ would take maximum at κ_{**} such that

$$\beta(u_{\kappa**}^*; N) = \beta. \tag{6.8}$$

This can be shown by taking derivative of $\zeta_\kappa(\beta; V)$ of Eq. (6.6) as

$$\frac{\partial}{\partial\kappa}\zeta_\kappa(\beta; V) = \left[\beta(u_\kappa^*; N) - \frac{2\kappa}{(l - u_\kappa^*)}\right]\frac{\partial u_\kappa^*}{\partial\kappa} + 2\ln\left[\beta\Big/\frac{2\kappa}{(l - u_\kappa^*)}\right] \tag{6.9}$$

$$= 2\ln\left[\frac{\beta}{\beta(u_\kappa^*; N)}\right]. \tag{6.10}$$

As κ is increased, u_κ^* decreases and $\beta(u_\kappa^*; N)$ increases. Therefore, with increasing κ, this derivative decreases monotonically from positive values (for $\kappa < \kappa_{**}$) to zero (at $\kappa = \kappa_{**}$) and to negative values (for $\kappa > \kappa_{**}$). Hence, R_k takes maximum at $\kappa = \kappa_{**}$, if κ takes continuous values.

Although κ actually takes discrete values ($= 0, 1/N, 2/N, \cdots$), I can find a value(s) κ_* among these values such that $\left|\frac{\partial}{\partial\kappa}\zeta_\kappa(\beta; V)\right|$ is minimum. u_κ also takes discrete values, whose intervals are

$$u_{\kappa\pm1/N} - u_\kappa = \Theta(1/N). \tag{6.11}$$

Hence, from Eq. (4.18),

$$u_{\kappa\pm1/N}^* - u_\kappa^* = \Theta(1/N). \tag{6.12}$$

Therefore,

$$u_{\kappa*}^* - u_{\kappa_{**}}^* = \Theta(1/N), \tag{6.13}$$

$$\beta(u_{\kappa_*}^*; N) = \beta + \Theta(1/N). \tag{6.14}$$

Since $\beta(u; V) = \Theta(1)$, there exists a constant $\Delta\kappa$ of $\Theta(1)$ such that

$$\frac{\partial}{\partial \kappa}\zeta_\kappa(\beta; V) = \begin{cases} \Theta(1)(\kappa \le \kappa_* - \Delta\kappa) \\ -\Theta(1)(\kappa \ge \kappa_* + \Delta\kappa) \end{cases} \tag{6.15}$$

Therefore, there exists a positive constant η of $\Theta(1)$ such that

$$\zeta_{\kappa_*}(\beta; V) - \zeta_\kappa(\beta; V) \ge \eta|\kappa - \kappa_*| \quad \text{for} |\kappa - \kappa_*| \ge \Delta\kappa. \tag{6.16}$$

Hence, for $\kappa - \kappa_* \ge \Delta\kappa$, I have the asymptotic inequality

$$|R_k| \le \exp\left[\frac{N}{2}\zeta_{\kappa_*}(\beta; V)\right] \exp\left[-\frac{N}{2}\eta(\kappa - \kappa_*)\right] \tag{6.17}$$

$$= \exp\left[\frac{N}{2}\zeta_{\kappa_*}(\beta; V)\right] \exp\left[-\frac{\eta}{2}(k - k_*)\right], \tag{6.18}$$

where $k_* \equiv N\kappa_*$. I take $k_{\max}$ arbitrarily such that

$$k_{\max} \ge k_* + N\Delta\kappa. \tag{6.19}$$

Then I have

$$\left\| \sum_{k \ge k_{\max}} R_k |\psi_k\rangle \right\| \le \sum_{k \ge k_{\max}} R_k \tag{6.20}$$

$$\le \exp\left[\frac{N}{2}\zeta_{\kappa_*}(\beta; V)\right] \exp\left[\frac{\eta}{2}k_*\right] \sum_{k \ge k_{\max}} \exp\left[-\frac{\eta}{2}k\right] \tag{6.21}$$

$$= \exp\left[\frac{N}{2}\zeta_{\kappa_*}(\beta; V)\right] \exp\left[\frac{\eta}{2}(k_* - k_{\max})\right] \frac{1}{1 - e^{-\eta/2}}, \tag{6.22}$$

which vanishes as $k_{\max} \to \infty$. Since k_* is a function of β and $k_{\max} > k_*$, this means that the series converges quickly for each value of β. Therefore, if I take arbitrarily the upper bound $\beta_{\max}$ of β, then the series converges uniformly for all β such that $0 < \beta \le \beta_{\max}$.

Because of this good convergence, I can obtain inversely the mTPQ state from the canonical one, e.g., by

$$|k\rangle = \left(\frac{2}{N}\right)^k \frac{\partial^k}{\partial \beta^k} e^{V\beta l/2} |\beta, V\rangle \bigg|_{\beta=0}. \tag{6.23}$$

The gTPQ state is obviously decomposed into the cTPQ states.

$$|\beta, \nu, V\rangle = \sum_N e^{\beta\nu N} |\beta, N, V\rangle \tag{6.24}$$

where $|\beta, N, V\rangle$ is the cTPQ state whose number of particles is restricted to N.

6.2 Practical Formula

Using Eqs. (6.3) and (6.24), one can perform the numerical calculation by just generating the mTPQ states. The mTPQ state can be generated by multiplying $(l - \hat{h})$ with $|\psi_0\rangle$ k times. Since R_k has a sharp peak at k^* (given by Eq. (6.8)), one can terminate the sum at a finite number k_{term}. It is sufficient to take k_{term} such that $k_{\text{term}} - k^*_{\max} = \Theta(N)$, where $k^*_{\max}$ is k^* corresponding to $\beta_{\max}$. Since I can show that $k^* = \Theta(N)$ for any finite β, $k^*_{\max} = \Theta(N)$. Hence, $k_{\text{term}} = \Theta(N)$. In this way, one can obtain $|\beta, V\rangle$ by multiplying $(l - \hat{h})$ repeatedly $\Theta(N)$ times. In this procedure, however, he can only get the cTPQ (or gTPQ) state for the fixed temperature β. Thus, he has to repeatedly generate the cTPQ states for different temperatures.

In order to overcome this inefficiency, I introduce another formula which is useful in practical applications. In fact, I use this formula to the applications in Chap. 7. All macroscopic variables can be calculated from the obtained $|\beta, V\rangle$. One can also calculate them without obtaining $|\beta, V\rangle$ explicitly. To show this, we evaluate $\langle\beta, V|\hat{A}|\beta, V\rangle$ and $\langle\beta, V|\beta, V\rangle$. Since the latter is included in the former as the case of $\hat{A} = \hat{1}$, I consider the former. From Eq. (6.3),

$$\langle\beta, V|\hat{A}|\beta, V\rangle = e^{-V\beta l} \sum_{k,k'} \frac{1}{k!k'!} \left(\frac{V\beta}{2}\right)^{k+k'} \langle k|\hat{A}|k'\rangle \tag{6.25}$$

In the case of $[\hat{A}, \hat{h}] \neq 0$, one has to calculate all combinations of $\langle k|\hat{A}|k'\rangle$, whose number of the combinations is $\Theta(N^2)$. For the special case where $[\hat{A}, \hat{h}] = 0$, this reduces to

$$\langle\beta, V|\hat{A}|\beta, V\rangle = \{\hat{A}\}'_{\beta,V}, \tag{6.26}$$

where

$$\{\hat{A}\}'_{\beta,V} \equiv \sum_{k=0}^{\infty} \frac{(V\beta)^{2k}}{(2k)!} \langle k|\hat{A}|k\rangle + \sum_{k=0}^{\infty} \frac{(V\beta)^{2k+1}}{(2k+1)!} \langle k|\hat{A}|k+1\rangle. \tag{6.27}$$

Even when $[\hat{A}, \hat{h}] \neq 0$, I can prove that Eq. (6.26) holds extremely well. Specifically, for

$$\{\hat{A}\}^{\text{TPQ}}_{\beta,V} \equiv \frac{\{\hat{A}\}'_{\beta,V}}{\{\hat{1}\}'_{\beta,V}} \tag{6.28}$$

and

$$u_n(A)^2 \equiv \overline{(\{\hat{A}\}^{\text{TPQ}}_{\beta,V} - \langle\hat{A}\rangle^{\text{ens}}_{\beta,V})^2}, \tag{6.29}$$

I have

$$\mathrm{P}\left(\left|\{\hat{A}\}^{\mathrm{TPQ}}_{\beta,V} - \langle \hat{A}\rangle^{\mathrm{ens}}_{\beta,V}\right| \geq \epsilon\right) \leq u_n(A)^2/\epsilon^2, \tag{6.30}$$

$$u_n(A)^2 \leq \frac{\langle (\Delta\hat{A})^2\rangle^{\mathrm{ens}}_{\beta,V}}{\epsilon^2 \exp[V\beta\{f(0;V) - f(1/\beta;V)\}]}. \tag{6.31}$$

Eq. (6.31) means that

$$\{\hat{A}\}^{\mathrm{TPQ}}_{\beta,V} \xrightarrow{P} \langle \hat{A}\rangle^{\mathrm{ens}}_{\beta,V} \tag{6.32}$$

exponentially fast and uniformly. Formula (6.26) is more useful than Formula (6.25) because one needs only to calculate $\langle k|\hat{A}|k\rangle$ and $\langle k|\hat{A}|k+1\rangle$ for all $k \leq k_{\mathrm{term}}$ to obtain the results for *all* $\beta \leq \beta_{\max}$.

6.3 The TPQ Formulation of Statistical Mechanics

To sum up the results which have been obtained, I have established the new formulation of statistical mechanics, in the same level as the ensemble formulation. The TPQ formulation, which consists of three types of the TPQ state, the microcanonical, canonical, and grandcanonical TPQ states, is summarized as follows. Depending on the choice of independent variables, (E, N) or (β, N), one can use either state, because they give identical thermodynamic results. A single realization of a TPQ state suffices for evaluating all quantities of statistical-mechanical interest. Moreover, one can estimate the upper bounds of errors (which vanish as $N \to \infty$) by formulas (3.14), (3.18), (6.30), and (6.31). The microcanonical and canonical TPQ states are transformed into each other by simple analytic relations, Eqs. (6.3) and (6.23). Hence, getting either one implies getting both. Using this fact, I have developed a practical formula (6.26). In addition to this, it is straightforward to extend these results to the TPQ states corresponding to other ensembles, such as the grand canonical ensemble.

Regarding the choice between the canonical and microcanonical TPQ states, one can use either depending on the purpose. For example, if one is interested in a first-order phase transition at which the specific heat $c = \partial u/\partial T = (\partial T/\partial u)^{-1}$ diverges the microcanonical one is practically better, because $T(u)$ is continuous (whereas $u(T)$ is discontinuous) through the transition [2, 3]. On the other hand, the canonical one is better when one studies low-temperature behavior of c, because $\partial u(T)/\partial T$ gets small ($\partial T(u)/\partial u$ diverges) as $c \to 0$.

References

1. S. Sugiura, A. Shimizu, Phys. Rev. Lett. **111**, 010401 (2013)
2. H.B. Callen, *Thermodynamics* (John Wiley and Sons, New York, 1960)
3. A. Shimizu, J. Phys. Soc. Jpn. **77**, 104001 (2008)

Chapter 7
Application to Numerical Calculation

In this chapter, I survey the TPQ formulation in terms of a numerical method. Firstly, I review related methods for quantum statistical mechanics and compare them to my formulation. Then, I show some results on the numerics using the TPQ formulation for the frustrated quantum spin system. Our results are performed for the largest number of spins ever in that model.

7.1 TPQ Formulation as Numerical Method

In the TPQ formulation, the mixed state which represents the equilibrium state is replaced by a pure quantum state. This formulation is not only interesting for the foundation of statistical mechanics, but also useful to a numerical method. The merits of the TPQ formulation comparing to other numerical methods are listed below:

1. Single realization is sufficient
   Calculating the thermal equilibrium values, there have been many methods in which one samples a small number of pure states. To the best of my knowledge, however, the TPQ formulation is the first formulation in which the error is concretely evaluated and revealed to be exponentially small. These findings guarantee that a single realization is sufficient to evaluate all physical quantities of statistical mechanical interest, which include the genuine thermodynamic variables and dynamical quantities such as the time correlation functions.
2. No limitation of models
   Since the TPQ states can be generated only by multiplying Hamiltonian matrix to a vector, there are no restrictions of models or spatial dimensions. The TPQ

S. Sugiura, *Formulation of Statistical Mechanics Based on Thermal Pure Quantum States*, Springer Theses, DOI 10.1007/978-981-10-1506-9_7

states have been applied to spin-1/2 Kagome Heisenberg antiferromagnet [12] and Hubbard model on a triangular lattice [1], which are well-known as frustrated systems.

3. Easy to perform and parallelize
   In the numerical calculation using microcanonical, canonical, or grandcanonical TPQ states, a computer program consists of the generation of a random initial vector and the multiplication of Hamiltonian matrix to the vector. Not only this program is simple to code, but also this simpleness allows us to utilize library codes and efficient parallelization. Especially, Hamiltonian used in statistical mechanics is a sparse matrix. Thus, the parallelization efficiency is very high.
4. Formulation for finite temperature
   Since there are an exponentially large number of relevant states at finite temperature, it is difficult to know the exact behavior of a physical system at finite temperature. In fact, only several integrable models have been solved exactly at finite temperature, while ground states have been obtained in a lot of models. In the numerical simulation, I consider DMRG as an example. When DMRG is applied to the finite temperature calculation, the number of the kept basis is exponentially increases, although the number of it is $O(\text{Poly}(V))$ at the ground state. By contrast, as I have explicitly shown in formulas (3.14), (3.18), and (6.31), the accuracy of the TPQ states gets better as the number of the relevant states increases. The TPQ states take advantage of the dense number of states to make the calculation accurate.
5. Almost self-validating
   In Formulas (3.14), (3.18), and (6.31), all the terms in the right hand sides can be evaluated by the TPQ states themselves. Therefore, one can estimate the upper bound of errors from these formulas. This estimation will be practiced later.
6. Need to store only two vectors
   Owing to formula (6.26), the calculation can be performed by keeping the last state $|k\rangle$ in a computer memory and generating $|k+1\rangle$. One can throw $|1\rangle, \cdots, |k-1\rangle$ away from the memory.
7. Free from orthogonality among states
   In numerical calculations such as Lanczos method, the orthogonality among basis states often debase the numerical accuracy. In the Lanczos method, one has to reorthogonalize all the basis states when the number of states exceed around 100. By contrast, the orthogonality is never imposed on the mTPQ states $|1\rangle, \cdots, |k\rangle$. Owing to it, one can easily perform the numerical calculation for $k > 10^3$.

7.2 Comparison to Other Methods

In the viewpoint of the numerical calculation method, there are tens of competitive methods. Among them, I compare the TPQ states to three major methods and three similar methods. The former three are quantum Monte Carlo method (QMC) [2], density matrix renormalization group (DMRG) [3], and numerical diagonalization (ND). The latter three are quantum transfer Monte Carlo method (QTMC) [8], finite temperature Lanczos method (FTLM) [9], and Kernel polynomial method (KPM) [10].

7.2.1 *Quantum Monte Carlo*

Quantum Monte Carlo method (QMC) [2] is one of the most successful methods in statistical mechanics. In QMC, a path integral is mapped onto a classical system, and a small number of paths are sampled stochastically. When it works, its numerical cost is $O(\mathrm{Poly}(V))$, which is far better than the TPQ states. However, QMC has an infamous "sign problem," which appears in the frustrated system and the fermion system. The failure of QMC in these models is because the quantum state in these models at low temperature cannot be represented by the sum of the polynomial order of classical states. Thus, the TPQ states have an advantage in the point 2 of the previous section. It is also advantageous in point 3.

7.2.2 *Density Matrix Renormalization Group*

Density matrix renormalization group (DMRG) [3] has been one of the hottest numerical methods recently. It represents a quantum state by a matrix product state. When this representation is efficient, one can reduce the dimension of the matrix χ by truncating many of irrelevant states. In 1D system, $\chi = O(\mathrm{Poly}(V))$, which is very efficient, is enough to describe the ground state. By contrast, in d (≥ 2) dimensional system, $\chi = O(\exp[V])$ is needed. Thus, many people had been considering that DMRG is the method only for 1D. However, after the difficulty of the frustrated system and the fermion system, e.g., Kagome lattice and Hubbard model have been recognized, DMRG is reconsidered as the method for 2Ds. It can simulate the ground state for Kagome lattice with the lattice site $N > 100$. There are also suggested a lot of related methods, which are collectively called tensor network states. For example, multiple entanglement renormalization ansatz (MERA) [4] and projected entangled pair states(PEPS) [5] are famous among them. However, the numerical procedure of the tensor network states is not easy to be programmed.

For finite temperature, imaginary time evolution of a pure quantum state [6] is often used. Since the thermo field dynamics (TFD) is compatible to DMRG, it has been widely used. Furthermore, minimally entangled typical states (METTs) were proposed [7]. Unlike the TFD, the METTs do not need any ancilla. The latter seems to be better than the former and is similar to the TPQ states. However, the state generated in METTs is the state $|i; \beta, V\rangle$, which is defined in Eq. (3.2). $|i; \beta, V\rangle$ cannot be the TPQ state. Returning to the numerical efficiency of DMRG, the entanglement gets exponentially larger as temperature is lowered in the DMRG calculation. To describe the state with the exponentially large entanglement, exponentially large χ is needed. This is the bottleneck of DMRG.

To sum up, points 1, 2, 3, and 4 are the advantages of the TPQ states. Among them, points 2 and 4 are important. If one is interested in the ground state or 1D system, DMRG will be the first choice for him.

7.2.3 Numerical Diagonalization

Numerical diagonalization (ND) of a matrix is a robust method, which is used in many fields in science. In statistical mechanics, we usually diagonalize a Hamiltonian matrix and obtain energy eigenvalue (and energy eigenstates) numerically. For the ground state, Lanczos method is often employed. The numerical cost of the Lanczos method is the same order as the TPQ state. At finite temperature, however, one needs to obtain a lot of energy eigenvalues to calculate the ensemble average. In this case, the orthogonality among the basis state becomes a serious problem in the Lanczos method. Instead of it, other ND methods such as Hausdorff method are employed. However, its numerical cost is more than square of that of the TPQ state. Therefore, the TPQ state is far better than ND at finite temperature.

7.2.4 Transfer Matrix Monte Carlo Method

In 1987, Imada and Takahashi proposed transfer matrix Monte Carlo method (tMC). It uses the same state as the cTPQ state defined in Eq. (3.8). In the practical calculation, one starts a calculation by preparing random vector $|\psi_0\rangle$. Then, one multiplies $\exp[-\frac{1}{2}\beta\hat{H}]$ to it. Up to here, tMC is the same as the cTPQ state in terms of numerical methods. In the tMC, however, this state is regarded as just a sample state drawn from the Gibbs state. Then, the calculation is justified by the sample average over many realizations of this state, and the error of the calculation is estimated by the sample variance. Hence, the theoretical aspect is completely different from the TPQ formulation. In the TPQ formulation, the error is upper bounded in good accuracy, and it is guaranteed that the error is exponentially small (although the inventors of tMC realized that the error is very small). As mentioned in the point 5, these formula enable us to estimate the error by the TPQ state itself.

Aside from the theoretical aspects, when we compare the tMC and the cTPQ state in terms of numerical methods, they have many points in common. Both of the calculations based on the multiplication of Hamiltonian, although the way to multiplication of $\exp[-\frac{1}{2}\beta\hat{H}]$ is different. In tMC, one uses transfer matrix technique. In the cTPQ, one uses the mTPQ state. The numerical costs of both methods are also similar. Among them, a clear advantage of the cTPQ state is the formula (6.31). This enables us to get the result for all β continuously by calculating a sequence of the mTPQ states once.

7.2.5 Finite Temperature Lanczos Method

Finite temperature Lanczos method (FTLM) is a method based on the Lanczos technique and high-temperature expansion. Since our formula (6.5) is also the high-temperature expansion, the FTLM and the numerical calculation of the cTPQ state are similar. The difference is that the FTLM employs the Lanczos technique, and the cTPQ state uses the decomposition of the mTPQ state. Both Lanczos technique and the generation of the mTPQ state are performed by multiplying Hamiltonian to a vector, the numerical cost is similar. However, the Lanczos technique has the difficulty of the orthogonality.

In the Lanczos technique, one considers the Krylov subspace,

$$\mathscr{K} \equiv \mathrm{span}\{|v\rangle, \hat{A}|v\rangle, \hat{A}^2|v\rangle, \cdots, \hat{A}^k|v\rangle\}, \tag{7.1}$$

where b is an initial vector, and $\hat{A}$ is a matrix, as which we choose $\hat{H}$ now. Then, one prepares an orthogonal basis set spanning $\mathscr{K}$ and diagonalizes $\hat{A}$ only within $\mathscr{K}$. Using the eigenvalues and the eigenstates which are obtained by this diagonalization, the FTLM evaluates the terms in the high-temperature expansion. However, the orthogonality among the basis in $\mathscr{K}$ gets inaccurate as k increases, because of the round-off errors on a computer. Practically, when $k > 100$, it becomes a severe problem.

This k is directly related to the degree of the high-temperature expansion. So is the k of the mTPQ state. In the FTLM, k is limited around $k \sim 100$. When k exceed 100, the calculation cost increases a lot because the reorthonormalization is needed. In the mTPQ state, as I will demonstrate in the frustrated spin model, one can easily take $k > 1000$ because of the absence of the requirement of the orthogonality. Therefore, the point 7 seems to be an advantage of the TPQ state comparing to the FTLM. In addition to this, one has to keep all the basis in $\mathscr{K}$ in usual FTLM calculations while the cTPQ state needs to keep the last mTPQ state $|k\rangle$ only. The point 6 is also the advantage.

7.2.6 Kernel Polynomial Method

Kernel polynomial method is the general name for the methods which expand a function by some orthogonal polynomials. Among them, Chebyshev polynomials expansion [11] is often used in condensed matter physics. The computational cost is the same order as the TPQ state.

7.3 Application to Kagome Lattice Model

Using the TPQ formulation, I study the spin-1/2 Kagome Heisenberg antiferromagnet (KHA) [12]. KHA is a frustrated 2D quantum spin system. The KHA is expected to have a spin liquid state in the ground state. Calculating the KHA, QMC, which is one of the most efficient methods for quantum simulation, has the negative sign problem. The problem gets worse as temperature is lowered. Thus, QMC is difficult to be used to the KHA. Using ND, one gets the ground state for finite size. The KHA up to the number of sites $N = 42$ is analyzed by ND on supercomputers [13]. Among these numerical methods, the most successful method for the KHA is the tensor network calculations including DMRG and MERA. In the studies using DMRG, systems of over 100 sites were analyzed [14]. The results of these studies show the strong evidence that the ground state is a Z_2 spin liquid [14, 15].

When I shift our vision to finite temperature, it was suggested that the specific heat c has double peaks at low temperature [16]. However, the problem is still in dispute due to the complexity of the frustration and the finite size effect [16–20]. QMC still has a difficulty of the negative sign problem. In ND, one has to know the energy spectrum of all energy eigenstates in order to analyze at finite temperature. Hence, ND is only available up to around $N = 20$. There are some DMRG-based methods for finite temperature such as METTs. However, these methods require more numerical cost in the analysis at finite temperature than in the analysis of the ground state, because the number of excitation states increases exponentially and one has to retain these states. After reviewing these circumstances, it turns out that the TPQ states seem to be suitable for the analysis of the KHA, because the analysis for the large cluster has not be performed yet at finite temperature.

I apply the TPQ formulation to the KHA with $N = 18$-30. I compute specific heat c, free energy density f, and entropy density s. I employ formula (6.27). In formula (6.27), I terminate the sum up to $k_{\mathrm{term}} = 2000$, where series expansion has converged so that the residual is evaluated to be less than $10^{-10}\%$ for $T \geq 0.02J$.

In Fig. 7.2 I plot c, which is obtained using

$$\frac{\partial \langle \hat{h} \rangle^{\mathrm{ens}}_{\beta,V}}{\partial \beta} = -\langle (\hat{h} - \langle \hat{h} \rangle^{\mathrm{ens}}_{\beta,V})^2 \rangle^{\mathrm{ens}}_{\beta,V} \tag{7.2}$$

$$\simeq \{\hat{\Delta}^2\}^{\mathrm{TPQ}}_{\beta,V}, \tag{7.3}$$

where $\hat{\Delta} \equiv \hat{h} - \{\hat{h}\}^{\mathrm{TPQ}}_{\beta,V}$. I have also calculated c in the difference way as

$$\frac{\partial \langle \hat{h} \rangle^{\mathrm{ens}}_{\beta,V}}{\partial \beta} \simeq \frac{(\{\hat{h}\}^{\mathrm{TPQ}}_{\beta+\delta\beta,V} - \{\hat{h}\}^{\mathrm{TPQ}}_{\beta,V})}{\delta\beta}. \tag{7.4}$$

The difference of these two numerical results is much smaller than the line width of the data in Fig. 7.2. For $N = 27$ and 30, for which ND has never been performed, there is not a peak but a shoulder around $T = 0.1J$. Although the finite size effect

Fig. 7.1 f and s versus T for N = 30

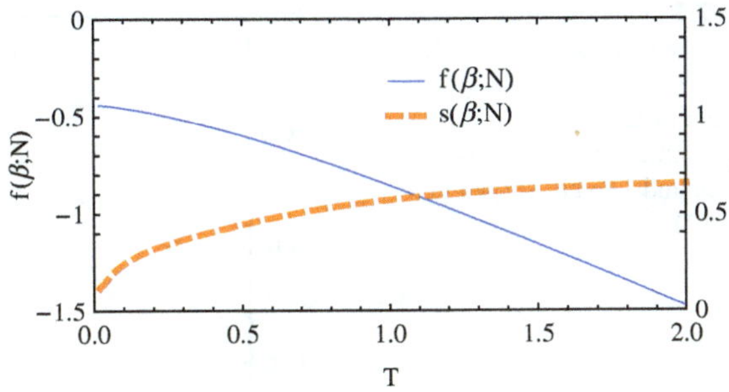

may still be non-negligible, these results strongly suggest that the peak of c at lower temperature will vanish in $N \to \infty$

I also estimate the error caused from the random initial vector $|\psi_0\rangle$ by using inequality (6.31). In Fig. 7.1, I plot f (left scale), which are calculated from Eq. (3.10). Using the results for f and those for

$$\{(\hat{\Delta}^2 - \{\hat{\Delta}^2\}_{\beta,V}^{\mathrm{TPQ}})^2\}_{\beta,V}^{\mathrm{TPQ}}, \tag{7.5}$$

I find that the (normalized) standard deviation

$$\frac{D_N(\hat{\Delta}^2)}{\{\hat{\Delta}^2\}_{\beta,V}^{\mathrm{TPQ}}} \tag{7.6}$$

for $N = 30$ is less than 1% down to $T = 0.1J$. The error of f itself is also estimated to be less than 1% down to $T = 0.1J$. In fact, when I estimate the standard deviation by generating tens of realizations of the cTPQ state, it is less than 1% down to $T = 0.1J$. Such a small error is attained because our method gets more accurate for larger entropy Ns, and the KHA has relatively large s at low temperature due to the frustration effect [18]. To see this quantitatively for $N = 30$, I plot s in the inset of Fig. 7.1 (right scale). Since the norm of the cTPQ state gives free energy, I convert it into entropy by Legendre transformation,

$$s = (u - f)\beta. \tag{7.7}$$

At $T = 0.2J$ there remains 45% of the total entropy ($= N \ln 2$). Such a large entropy makes $D_N(\hat{\Delta}^2)$ small.

Finally, to confirm the validity, I compute c for $N = 18$, for which the result of the ND is available [17]. For such a small cluster, the standard deviation estimated from inequality (6.31) is about 35% at $T = 0.1J$. Hence, I have taken average over 100 realizations of the TPQ state for $N = 18$ only. The difference between our results (18a, 18b) and those by the ND [17] is less than the line width of the data in Fig. 7.2.

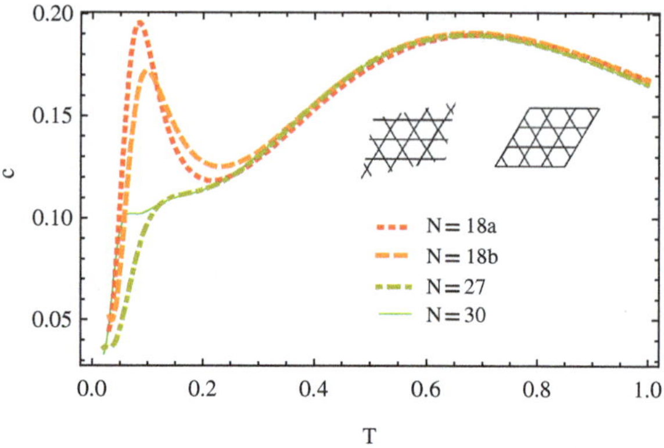

Fig. 7.2 Specific heat c versus T of the KHA for N = 18–30. (*Inset*) The shapes of the clusters of N = 30 (*left*) and 27 (*right*). Those of 18a and 18b are shown in Ref. [21]

References

1. M. Hyuga, S. Sugiura, K. Sakai, A. Shimizu, Phys. Rev. B **90**, 121110(R) (2014)
2. M. Suzuki, *Quantum Monte Carlo Methods in Condensed Matter Physics* (World Scientific, Singapore, 1993). editor
3. S.R. White, Phys. Rev. Lett. **69**, 2863 (1992)
4. V. Vidal, Phys. Rev. Lett. **99**, 220405 (2007)
5. V. Murg, F. Verstraete, J.I. Cirac, Phys. Rev. A **75**, 033605 (2007)
6. S.R. White, A.E. Feiguin, Phys. Rev. Lett. **93**, 076401 (2004)
7. S.R. White, Phys. Rev. Lett. **102**, 190601 (2009)
8. M. Imada, M. Takahashi, J. Phys. Soc. Jpn. **55**, 3354 (1986)
9. J. Jaklič, P. Prelovšek, Adv. Phys. **49**, 1 (2000)
10. A. Weiße, G. Wellein, A. Alvermann, H. Fehske, Rev. Mod. Phys. **78**, 275 (2006)
11. T.J. Rivlin, *Chebyshev Polynomials: From Approximation Theory to Algebra and Number Theory, Pure and Applied Mathematics*, 2nd edn. (Wiley, New York, 1990)
12. S. Sugiura, A. Shimizu, Phys. Rev. Lett. **111**, 010401 (2013)
13. H. Nakano, T. Sakai, J. Phys. Soc. Jpn. **83**, 104710 (2014)
14. S. Depenbrock, I.P. McCulloch, Phys. Rev. Lett. **109**, 067201 (2012)
15. M. Punk, D. Chowdhury, S. Sachdev, Nat. Phys. **10**, 289 (2014)
16. V. Elser, Phys. Rev. Lett. **62**, 2405 (1989)
17. N. Elstner, A.P. Young, Phys. Rev. B **50**, 6871 (1994)
18. P. Sindzingre, G. Misguich, C. Lhuillier, B. Bernu, L. Pierre, Ch. Waldtmann, H.-U. Everts, Phys. Rev. Lett. **84**, 2953 (2000)
19. M. Isoda, H. Nakano, T. Sakai, J. Phys. Soc. Jpn **80**, 084704 (2011)
20. H. Kawamura, K. Watanabe, T. Shimokawa, J. Phys. Soc. Jpn. **83**, 103704 (2014)
21. A.M. Läuchli, J. Sudan, E.S. Sørensen, Phys. Rev. B **83**, 212401 (2011)

Chapter 8
Conclusion

In this thesis, I have established the new formulation of statistical mechanics based on the TPQ state. Moreover, although the TPQ formulation give all the predictions of statistical-mechanical interest correctly, I have examined its difference from the ensemble formulation by stepping in the microscopic world. Then, in order to make the most of the advantage of the TPQ formulation, I have derived the formulas to transform from the TPQ states into the others. Owing to these formulas, I applied the TPQ formulation to the frustrated quantum spin model.

In Chap. 3, I generally defined pure quantum states that can represent an equilibrium state the TPQ states. Then, I introduced the cTPQ state, i.e., the TPQ state which is specified by temperature. Not only the mechanical variables but also the genuine thermodynamic variables are obtained from a single realization of the cTPQ state. In particular, the free energy is obtained from the norm of the cTPQ state, within the exponentially small error. Any equilibrium value of the MV can be obtained by the expectation value using the cTPQ state within the exponentially small error. Once we get both the free energy and any equilibrium values of the MVs, we can get all the predictions in statistical mechanics from them. Therefore, the new formulation of statistical mechanics based on the cTPQ state is established. In addition to this, it is straightforward to introduce the gTPQ state, which corresponds to the grandcanonical ensemble.

In Chap. 4, I introduced the mTPQ state, one of the TPQ states which are specified by energy, and saw its properties. The mTPQ state focuses on practical applications. It can be generated easily by multiplying Hamiltonian to an initial vector. In the same manner as the cTPQ state, the norm of the mTPQ state gives entropy and the expectation value of the mechanical variable gives the microcanonical ensemble average. Here, the errors are also exponentially small. When we compare the mTPQ states to the other TPQ states, the most suitable one depends on the situation. For example, when one analyzes the first-order phase transition, the mTPQ state will be most suitable.

In Chap. 5, I compared the TPQ states to the equilibrium states in the conventional formulation using the ensembles. The TPQ state is macroscopically invariant under the time evolution. The TPQ state can also correctly give the prediction in linear

S. Sugiura, *Formulation of Statistical Mechanics Based on Thermal Pure Quantum States*, Springer Theses, DOI 10.1007/978-981-10-1506-9_8

response theory. Thus, the TPQ state and the conventional equilibrium state look as if they were completely the same state. However, when we examine the fluctuation of mechanical variables, a clue to understand the difference between them is obtained. In quantum mechanics, a state has a quantum fluctuation. Then, when we move to quantum statistical mechanics, a thermal fluctuation is added to the quantum one. Therefore, it is often explained that the fluctuation in quantum statistical mechanics is the sum of these two fluctuations. However, the equilibrium state in the ensemble formulation is a quantum mixed state, and the quantum mixed state has many ways to be decomposed into the classical mixture of pure quantum states. Because of this ununiqueness, the decomposition of the fluctuation into the quantum and the thermal ones cannot be determined uniquely. On the other hand, as the TPQ states are pure quantum states, they have the quantum fluctuation only. Namely, all the fluctuations, which are predicted by the ensemble average, are squeezed into the quantum fluctuation in the TPQ formulation. Thus, many microstates within the range of the fluctuation are superposed in the TPQ state. This superposing makes a single realization of the TPQ state possible to give all the statistical mechanical predictions including the fluctuation. Here, as we know that there are exponentially large number of energy eigenstates within the range of the energy fluctuation, the TPQ states seem to be the superposition of exponentially large number of states. This intuition is quantified by calculating quantum entanglement. By seeing the quantum entanglement, we clearly distinguish the difference between the TPQ state and the equilibrium state in the ensemble formulation. It is revealed that the entanglement of the TPQ states are large, and especially at high temperature, almost maximum. On the other hand, the entanglement of the conventional equilibrium states is small and vanishing to zero as temperature is raised. The entanglement of the TPQ states and that of the conventional ones are the extreme opposites. When we recall that the former states are pure and the latter ones are mixed, it is obvious that the appropriate measure can detect the difference. Since the quantum entanglement is not included in either the MVs or the GTVs, the entanglement is the quantity which is beyond statistical mechanics. Furthermore, it can detect the difference between the TPQ state and the conventional equilibrium state.

In Chap. 6, we derived the relations between the TPQ states. I gave the transforming formula from the mTPQ state into the cTPQ state and that from the cTPQ state into the gTPQ state. Through these formulas, we can represent the cTPQ and gTPQ states as the superposition of the mTPQ states. Moreover, I gave the transforming formula which is more suitable for practical applications as well. These results in this section enable us to carry out the efficient applications.

To make the most of the merit of the TPQ formulation, in Chap. 7, I performed a numerical calculation using the TPQ formulation. Since a single realization is sufficient in the TPQ formulation, it has an advantage over the ensemble average, which needs to calculate an average over all the realizable states. In particular, the TPQ formulation is efficient in 2D frustrated systems and fermion systems, because QMC and DMRG, which are two of the most powerful methods in condensed matter physics, become difficult to be performed. Therefore, I applied the cTPQ to the KHA, which is one of the most famous frustrated spin systems and revealed the new behavior of the specific heat.

Curriculam Vitae

Sho Sugiura

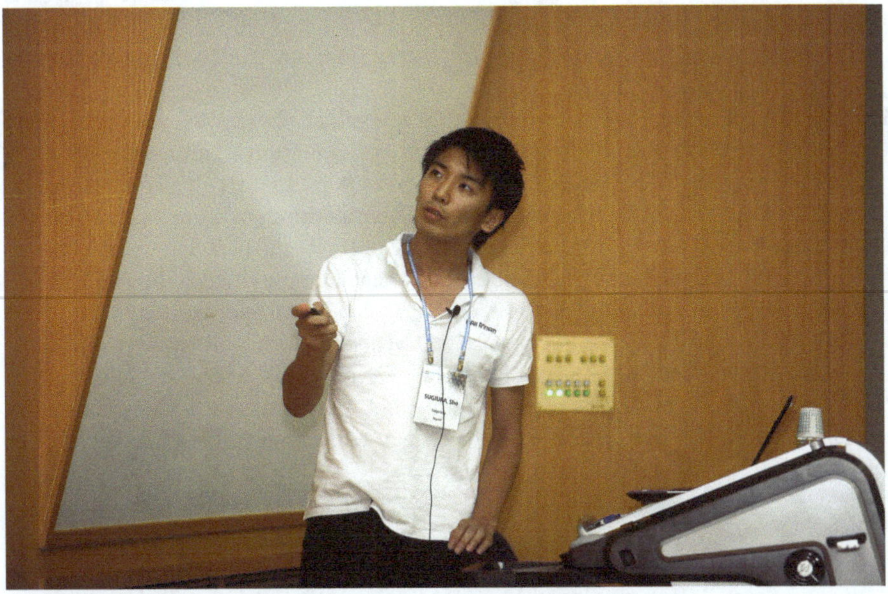

Address: University of Tokyo, Institute for Solid State Physics
5-1-5 Kashiwanoha, Kashiwa 277-8581 JAPAN
Phone: +81-80-3002-2386
Email: sugiura@issp.u-tokyo.ac.jp

Employment:

- 04/01/15-Current JSPS Postdoctoral Fellow, Institute for Solid State Physics, University of Tokyo

S. Sugiura, *Formulation of Statistical Mechanics Based on Thermal Pure Quantum States*, Springer Theses, DOI 10.1007/978-981-10-1506-9

Education:

- 2012–2015 **University of Tokyo**, Ph.D in Theoretical Physics (with honors)
  Thesis: Formulation of statistical mechanics based on thermal pure quantum state
  Advisor: Akira Shimizu
- 2010–2012 **University of Tokyo**, M.S. in Theoretical Physics
- 2006–2010 **University of Tokyo**, B.A. in Theoretical Physics

Awards:

- 5th JSPS Ikushi Prize, Mar. 2015 (to outstanding Ph.D candidates, awarded by the Emperor of Japan)
- 8th Young Scientist Award of the Physical Society of Japan, Mar. 2014

Publications:
Reserch Articles

1. SS and A.Shimizu, "Thermal pure quantum state at finite temperature", Physical Review Letters 108, 240401 (2012)
2. SS and A.Shimizu, "Canonical thermal pure quantum state", Physical Review Letters 111, 010401 (2013)
3. SS and A.Shimizu, "New formulation of statistical physics using thermal pure quantum states", Kinki University Series on Quantum Computing "Physics, Mathematics, and All that Quantum Jazz", Vol.9, p245 (2014),
4. M.Hyuga, SS, K.Sakai, and A.Shimizu, "Thermal pure quantum states of many-particle systems", Physical Review B 90, 121110(R) (2014)
5. S. Sasa, S. Sugiura, and Y. Yokokura, "Thermal pure state path integral and emergent symmetry", arXiv 1611.07268 (2016) (2) Introductory Articles
6. H. Fujita, Y. O. Nakagawa, S. Sugiura, and M. Watanabe, "Universality in volume law entanglement of pure quantum states", arXiv 1703.02993, (2017)
7. H. Fujita, Y. O. Nakagawa, S. Sugiura, and M. Oshikawa, "Construction of Hamiltonians by machine learning of energy and entanglement spectra", arXiv 1705.05372, (2017)

Introductory Articles

1. SS and A.Shimizu, "Formulation of statistical mechanics using thermal pure quantum state", Butsuri (Japan Physical Society Official Journal), Vol.70, Issue 5, p368 (2015)

Appendix A
Derivation of Formula (6.31)

In order to verify Formula (6.31), it is needed that both the variance of

$$\langle\psi_0|\hat{A}|\psi_0\rangle/\langle\psi_0|\hat{B}|\psi_0\rangle \tag{A.1}$$

is sufficiently small, where

$$\hat{A} \equiv \sum_k \frac{(V\beta)^k}{k!}(l-\hat{h})^{\lfloor k/2 \rfloor}\hat{M}(l-\hat{h})^{\lfloor (k+1)/2 \rfloor} \tag{A.2}$$

$$\hat{B} \equiv \exp(V\beta(l-\hat{h})). \tag{A.3}$$

In order to use Eqs. (3.43) and (3.45), what I need are the following three terms.

$$\begin{aligned}
&\overline{\left(\langle\psi_0|\hat{A}|\psi_0\rangle - \overline{\langle\psi_0|\hat{A}|\psi_0\rangle}\right)^2} \\
&= \sum_{n,k,k'} \frac{(V\beta)^{k+k'}}{k!k'!} \\
&\qquad \langle n|(l-\hat{h})^{\lfloor k/2 \rfloor}\hat{M}(l-\hat{h})^{\lfloor (k+1)/2 \rfloor}(l-\hat{h})^{\lfloor k'/2 \rfloor}\hat{M}(l-\hat{h})^{\lfloor (k'+1)/2 \rfloor}|n\rangle \\
&= \sum_{n,k,k'} \frac{(V\beta)^{k+k'}}{k!k'!} \\
&\qquad \langle n|(l-\hat{h})^{\lfloor k/2 \rfloor}\hat{M}(l-\hat{h})^{\lfloor (k+1)/2 \rfloor}(l-\hat{h})^{\lfloor k'/2 \rfloor}\hat{M}(l-\hat{h})^{\lfloor (k'+1)/2 \rfloor}|n\rangle
\end{aligned} \tag{A.4}$$

where $Z'(\beta) \equiv \sum_n \exp(V\beta(l-u_n))$)

S. Sugiura, *Formulation of Statistical Mechanics Based on Thermal Pure Quantum States*, Springer Theses, DOI 10.1007/978-981-10-1506-9

$$\overline{\left(\langle\psi_0|\hat{A}|\psi_0\rangle - \overline{\langle\psi_0|\hat{A}|\psi_0\rangle}\right)\left(\langle\psi_0|\hat{B}|\psi_0\rangle - \overline{\langle\psi_0|\hat{B}|\psi_0\rangle}\right)}$$
$$= \sum_{n,k} \frac{(V\beta)^k}{k!} \langle n|(l-\hat{h})^{\lfloor k/2\rfloor} \hat{M} (l-\hat{h})^{\lfloor (k+1)/2\rfloor} \exp(V\beta(l-\hat{h}))|n\rangle$$
$$= \sum_n \exp(2V\beta(l-u_n))\langle n|\hat{M}|n\rangle$$
$$= \langle\hat{M}\rangle^{\text{ens}}_{2\beta,V} Z'(2\beta) \tag{A.5}$$

$$\overline{\left(\langle\psi_0|\hat{B}|\psi_0\rangle - \overline{\langle\psi_0|\hat{B}|\psi_0\rangle}\right)^2} = \sum_n \langle n|\exp(2V\beta(l-\hat{h}))|n\rangle$$
$$= Z'(2\beta) \tag{A.6}$$

where I use Eq. (3.55). Using Eqs. (3.43), (A.5), and (A.6), I get the average.

$$\left|\overline{\langle\psi_0|\hat{A}|\psi_0\rangle/\langle\psi_0|\hat{B}|\psi_0\rangle} - \overline{\langle\psi_0|\hat{A}|\psi_0\rangle}/\overline{\langle\psi_0|\hat{B}|\psi_0\rangle}\right|$$
$$= \left| - \frac{\langle\hat{M}\rangle^{\text{ens}}_{2\beta,V} Z'(2\beta)}{Z'(\beta)^2} + \frac{Z'(2\beta)\langle\hat{M}\rangle^{\text{ens}}_{\beta,V}}{Z'(\beta)^2}\right| \tag{A.7}$$
$$= \frac{\left(\langle\hat{M}\rangle_{\text{can}}(2\beta) - \langle\hat{M}\rangle_{\text{can}}(\beta)\right)}{\exp[2V\beta(f(1/2\beta;V) - f(1/\beta;V))]} \tag{A.8}$$

The rhs of this equation is exponentially small. Using Eqs. (3.45), (A.4), (A.5), and (A.6), I get the variance

$$\overline{\left(\langle\psi_0|\hat{A}|\psi_0\rangle/\langle\psi_0|\hat{B}|\psi_0\rangle - \overline{\langle\psi_0|\hat{A}|\psi_0\rangle}/\overline{\langle\psi_0|\hat{B}|\psi_0\rangle}\right)^2}$$
$$= \sum_{n,m,k,k'} \frac{(V\beta)^{k+k'}}{k!k'!} (l-u_n)^{\lfloor k/2\rfloor}(l-u_m)^{\lfloor (k+1)/2\rfloor}(l-u_m)^{\lfloor k'/2\rfloor}(l-u_n)^{\lfloor (k'+1)/2\rfloor}|\langle n|\hat{M}|m\rangle|^2$$
$$/\left(Z'(\beta)\right)^2$$
$$-2\langle\hat{M}\rangle^{\text{ens}}_{2\beta,V} Z'(2\beta)\frac{\langle\hat{M}\rangle^{\text{ens}}_{\beta,V}}{\left(Z'(\beta)\right)^2}$$
$$+Z'(2\beta)\frac{\langle\hat{M}\rangle^{\text{ens}\,2}_{\beta,V}}{\left(Z'(\beta)\right)^2}$$
$$= \sum_{n,m\neq n,k,k'} \frac{(V\beta)^{k+k'}}{k!k'!} (l-u_n)^{\lfloor k/2\rfloor}(l-u_m)^{\lfloor (k+1)/2\rfloor}(l-u_m)^{\lfloor k'/2\rfloor}(l-u_n)^{\lfloor (k'+1)/2\rfloor}|\langle n|\hat{M}|m\rangle|^2/\left(Z'(\beta)\right)^2$$
$$+\sum_n \exp(2V\beta(l-u_n))|\langle n|\hat{M}|n\rangle|^2/\left(Z'(\beta)\right)^2$$
$$-2\left(\sum_n \exp(2V\beta(l-u_n))\langle n|\hat{M}|n\rangle\right)\langle\hat{M}\rangle^{\text{ens}}_{\beta,V}/\left(Z'(\beta)\right)^2$$

$$+\left(\sum_n \exp(2V\beta(l-u_n))\right)\langle\hat{M}\rangle_{\beta,V}^{\text{ens}\,2}/\left(Z'(\beta)\right)^2 \tag{A.9}$$

$$= Z'(\beta)^2$$

$$\left(\sum_{n,m\neq n,k,k'}\frac{(V\beta)^{k+k'}}{k!k'!}(l-u_n)^{\lfloor k/2\rfloor}(l-u_m)^{\lfloor (k+1)/2\rfloor}(l-u_m)^{\lfloor k'/2\rfloor}(l-u_n)^{\lfloor (k'+1)/2\rfloor}|\langle n|\hat{M}|m\rangle|^2\right.$$

$$\left.+\sum_n \exp(2V\beta(l-u_n))\left(\langle n|\hat{M}|n\rangle-\langle\hat{M}\rangle_{\beta,V}^{\text{ens}}\right)^2\right) \tag{A.10}$$

$$= Z'(\beta)^2$$

$$\left(\sum_{n,m\neq n,k,k'}\frac{(V\beta)^{k+k'}}{k!k'!}(l-u_n)^{\lfloor k/2\rfloor}(l-u_m)^{\lfloor (k+1)/2\rfloor}(l-u_m)^{\lfloor k'/2\rfloor}(l-u_n)^{\lfloor (k'+1)/2\rfloor}|\langle n|\hat{M}|m\rangle|^2\right.$$

$$\left.+\sum_{n,k,k'}\frac{(V\beta)^{k+k'}}{k!k'!}(l-u_n)^{\lfloor k/2\rfloor}(l-u_n)^{\lfloor (k+1)/2\rfloor}(l-u_n)^{\lfloor k'/2\rfloor}(l-u_n)^{\lfloor (k'+1)/2\rfloor}\left(\langle n|\hat{M}|n\rangle-\langle\hat{M}\rangle_{\beta,V}^{\text{ens}}\right)^2\right)$$

Here, 1st term and 2nd term are respectively positive. So, I get

$$\leq Z'(\beta)^2$$

$$\left(\sum_{n,m\neq n,k,k'}\frac{(V\beta)^{k+k'}}{k!k'!}(l-u_n)^{\lfloor k/2\rfloor}(l-u_{\min})^{\lfloor (k+1)/2\rfloor}\right.$$

$$(l-u_{\min})^{\lfloor k'/2\rfloor}(l-u_n)^{\lfloor (k'+1)/2\rfloor}|\langle n|\hat{M}|m\rangle|^2$$

$$+\sum_{n,k,k'}\frac{(V\beta)^{k+k'}}{k!k'!}(l-u_n)^{\lfloor k/2\rfloor}(l-u_{\min})^{\lfloor (k+1)/2\rfloor}(l-u_{\min})^{\lfloor k'/2\rfloor}(l-u_n)^{\lfloor (k'+1)/2\rfloor}$$

$$\left.\left(\langle n|\hat{M}|n\rangle-\langle\hat{M}\rangle_{\beta,V}^{\text{ens}}\right)^2\right)$$

$$= Z'(\beta)^2$$

$$\sum_{n,k,k'}\frac{(V\beta)^{k+k'}}{k!k'!}(l-u_n)^{\lfloor k/2\rfloor}(l-u_{\min})^{\lfloor (k+1)/2\rfloor}(l-u_{\min})^{\lfloor k'/2\rfloor}(l-u_n)^{\lfloor (k'+1)/2\rfloor}$$

$$\left(\left(\sum_{m\neq n}|\langle n|\hat{M}|m\rangle|^2\right)+\left(\langle n|\hat{M}|n\rangle-\langle\hat{M}\rangle_{\beta,V}^{\text{ens}}\right)^2\right) \tag{A.11}$$

$$= Z'(\beta)^2$$

$$\sum_{n,k,k'}\frac{(V\beta)^{k+k'}}{k!k'!}(l-u_n)^{\lfloor k/2\rfloor}(l-u_{\min})^{\lfloor (k+1)/2\rfloor}(l-u_{\min})^{\lfloor k'/2\rfloor}(l-u_n)^{\lfloor (k'+1)/2\rfloor}$$

$$\left(\left(\sum_m|\langle n|\hat{M}|m\rangle|^2\right)-2\langle n|\hat{M}|n\rangle\langle\hat{M}\rangle_{\beta,V}^{\text{ens}\,2}+\langle\hat{M}\rangle_{\beta,V}^{\text{ens}\,2}\right) \tag{A.12}$$

$$= Z'(\beta)^2\sum_n\left(\langle n|\hat{M}^2|n\rangle-2\langle n|\hat{M}|n\rangle\langle\hat{M}\rangle_{\beta,V}^{\text{ens}\,2}+\langle\hat{M}\rangle_{\beta,V}^{\text{ens}\,2}\right)$$

$$\sum_{k,k'}\frac{(V\beta)^{k+k'}}{k!k'!}(l-u_n)^{\lfloor k/2\rfloor}(l-u_{\min})^{\lfloor (k+1)/2\rfloor}(l-u_{\min})^{\lfloor k'/2\rfloor}(l-u_n)^{\lfloor (k'+1)/2\rfloor} \tag{A.13}$$

From here, I will prove $\sum_{k,k'}$ term in Eq. (A.13) satisfies

$$\sum_{k} \frac{(V\beta)^k}{k!}(l-u_n)^{\lfloor k/2 \rfloor}(l-u_{\min})^{\lfloor (k+1)/2 \rfloor} \sum_{k} \frac{(V\beta)^{k'}}{k'!}(l-u_{\min})^{\lfloor k'/2 \rfloor}(l-u_n)^{\lfloor (k'+1)/2 \rfloor}$$
$$\leq \exp(V\beta(l-u_n+l-u_{\min})) \tag{A.14}$$

I separate the sum of the l.h.s. of Eq. (A.14) in terms of even odd. Before calculate it, I calculate the following.

$$\sum_{k} \frac{(V\beta)^k}{k!}(l-e_a)^{\lfloor k/2 \rfloor}(l-e_b)^{\lfloor (k+1)/2 \rfloor}$$
$$= \sum_{k:\text{even}} \frac{(V\beta)^k}{k!}(l-e_a)^{k/2}(l-e_b)^{k/2} + \sum_{k:\text{odd}} \frac{(V\beta)^k}{k!}(l-e_a)^{(k-1)/2}(l-e_b)^{(k+1)/2} \tag{A.15}$$
$$= \sum_{k:\text{even}} \frac{(V\beta)^k}{k!}(l-e_a)^{k/2}(l-e_b)^{k/2} + \sqrt{\frac{(l-e_b)}{(l-e_a)}} \sum_{k:\text{odd}} \frac{(V\beta)^k}{k!}(l-e_a)^{k/2}(l-e_b)^{k/2} \tag{A.16}$$
$$= \cosh\left(V\beta\sqrt{(l-e_a)(l-e_b)}\right) + \sqrt{\frac{(l-e_b)}{(l-e_a)}} \sinh\left(V\beta\sqrt{(l-e_a)(l-e_b)}\right) \tag{A.17}$$

So, the sum of the l.h.s. of Eq. (A.14) becomes

$$\sum_{k} \frac{(V\beta)^k}{k!}(l-u_n)^{\lfloor k/2 \rfloor}(l-u_{\min})^{\lfloor (k+1)/2 \rfloor} \sum_{k} \frac{(V\beta)^{k'}}{k'!}(l-u_{\min})^{\lfloor k'/2 \rfloor}(l-u_n)^{\lfloor (k'+1)/2 \rfloor}$$
$$= \left(\cosh\left(V\beta\sqrt{(l-u_n)(l-u_{\min})}\right) + \sqrt{\frac{(l-u_{\min})}{(l-u_n)}} \sinh\left(V\beta\sqrt{(l-u_n)(l-u_{\min})}\right)\right)$$
$$\left(\cosh\left(hV\beta\sqrt{(l-u_{\min})(l-u_n)}\right) + \sqrt{\frac{(l-u_n)}{(l-u_{\min})}} \sinh\left(V\beta\sqrt{(l-u_{\min})(l-u_n)}\right)\right)$$
$$= \cosh^2\left(V\beta\sqrt{(l-u_n)(l-u_{\min})}\right) + \sinh^2\left(V\beta\sqrt{(l-u_n)(l-u_{\min})}\right)$$
$$+ \left(\sqrt{\frac{(l-u_{\min})}{(l-u_n)}} + \sqrt{\frac{(l-u_n)}{(l-u_{\min})}}\right)$$
$$\sinh\left(V\beta\sqrt{(l-u_n)(l-u_{\min})}\right) \cosh\left(V\beta\sqrt{(l-u_{\min})(l-u_n)}\right) \tag{A.18}$$
$$= \cosh\left(2V\beta\sqrt{(l-u_n)(l-u_{\min})}\right)$$
$$+ \frac{1}{2}\left(\sqrt{\frac{(l-u_{\min})}{(l-u_n)}} + \sqrt{\frac{(l-u_n)}{(l-u_{\min})}}\right) \sinh\left(2V\beta\sqrt{(l-u_n)(l-u_{\min})}\right) \tag{A.19}$$

Here, I notice that the first and second term are respectively positive. I can easily evaluate the first term as

$$\cosh\left(2V\beta\sqrt{(l-u_n)(l-u_{\min})}\right) = \sum_{k:\text{even}} \frac{(2V\beta)^k}{k!} \sqrt{(l-u_n)(l-u_{\min})}^k \tag{A.20}$$
$$\leq \sum_{k:\text{even}} \frac{(V\beta)^k}{k!}(l-u_n+l-u_{\min})^k \tag{A.21}$$

Here, I use the inequality of arithmetic and geometric mean. In the same manner, I can evaluate the second term as

$$\frac{1}{2}\left(\sqrt{\frac{(l-u_{\min})}{(l-u_n)}}+\sqrt{\frac{(l-u_n)}{(l-u_{\min})}}\right)\sinh\left(2V\beta\sqrt{(l-u_n)(l-u_{\min})}\right)$$
$$\leq \sum_{k:\text{odd}} \frac{(V\beta)^k}{k!}(l-u_n+l-u_{\min})^k \tag{A.22}$$

In order to prove Eq. (A.22), I expand the l.h.s of it.

$$\frac{1}{2}\left(\sqrt{\frac{(l-u_{\min})}{(l-u_n)}}+\sqrt{\frac{(l-u_n)}{(l-u_{\min})}}\right)\sinh\left(2V\beta\sqrt{(l-u_n)(l-u_{\min})}\right)$$
$$=\frac{1}{2}\left(\sqrt{\frac{(l-u_{\min})}{(l-u_n)}}+\sqrt{\frac{(l-u_n)}{(l-u_{\min})}}\right)\sum_{k:\text{odd}}\frac{(2V\beta)^k}{k!}\sqrt{(l-u_n)(l-u_{\min})}^{\,k}$$
$$=\frac{1}{2}\left((l-u_n)+(l-u_{\min})\right)\sum_{k:\text{odd}}\frac{(2V\beta)^k}{k!}\sqrt{(l-u_n)(l-u_{\min})}^{\,(k-1)}$$
$$\leq \sum_{k:\text{odd}} \frac{(V\beta)^k}{k!}(l-u_n+l-u_{\min})^k \tag{A.23}$$

Now, I have proved inequality (A.22). Using inequality (A.21) and (A.22), I get inequality (A.14). Then, I return the main proof. Using inequality (A.13), and (A.14), I get the final result

$$\overline{\left(\langle\psi_0|\hat{A}|\psi_0\rangle/\langle\psi_0|\hat{B}|\psi_0\rangle-\overline{\langle\psi_0|\hat{A}|\psi_0\rangle}/\overline{\langle\psi_0|\hat{B}|\psi_0\rangle}\right)^2}$$
$$\leq \frac{D}{(D+1)Z'(\beta)^2}\sum_n\left(\langle n|\hat{M}^2|n\rangle-2\langle n|\hat{M}|n\rangle\langle\hat{M}\rangle^{\text{ens}\,2}_{\beta,V}+\langle\hat{M}\rangle^{\text{ens}\,2}_{\beta,V}\right)$$
$$exp(V\beta(l-u_n+l-u_{\min}))$$
$$=\frac{D}{(D+1)}\frac{\exp(V\beta(l-u_{\min}))}{Z'(\beta)}\left(\langle\hat{M}^2\rangle^{\text{ens}}_{\beta,V}-\langle\hat{M}\rangle^{\text{ens}\,2}_{\beta,V}\right) \tag{A.24}$$
$$=\frac{D}{D+1}\frac{\langle(M-\langle M\rangle^{\text{ens}}_{\beta,V})^2\rangle^{\text{ens}}_{\beta,V}}{\exp[V\beta(f(0;V)-f(1/\beta;V))]} \tag{A.25}$$

So, I verify that the expectation values of mechanical variables can be calculated by using Eq. (A.1) at finite temperature.

Kolibri 360 GmbH
Ferdinand-Jühlke-Straße 7
99095 Erfurt, Deutschland
produktsicherheit@kolibri360.de